Cape Verdean Women and Globalization

Cape Verdean Women and Globalization

The Politics of Gender, Culture, and Resistance

Katherine Carter
and
Judy Aulette

palgrave
macmillan

CAPE VERDEAN WOMEN AND GLOBALIZATION

First published in 2009 by
PALGRAVE MACMILLAN®
in the United States—a division of St. Martin's Press LLC,
175 Fifth Avenue, New York, NY 10010.

Where this book is distributed in the UK, Europe and the rest of the world, this is by Palgrave Macmillan, a division of Macmillan Publishers Limited, registered in England, company number 785998, of Houndmills, Basingstoke, Hampshire RG21 6XS.

Palgrave Macmillan is the global academic imprint of the above companies and has companies and representatives throughout the world.

Palgrave® and Macmillan® are registered trademarks in the United States, the United Kingdom, Europe and other countries.

ISBN: 978–0–230–61808–4

Library of Congress Cataloging-in-Publication Data is available from the Library of Congress.

A catalogue record of the book is available from the British Library.

Design by Newgen Imaging Systems (P) Ltd., Chennai, India.

First edition: October 2009

10 9 8 7 6 5 4 3 2 1

Printed in the United States of America.

Contents

Preface and Acknowledgments

This book represents a long journey for the two authors who met more than a decade ago. Since then they have both worked and lived in several countries including Cape Verde, Hungary, England, Ethiopia, Iraq, Poland, Scotland, South Africa, and the United States. This study began when Katherine Carter landed in Cape Verde to teach English. Carter came to know the people in her community and school and she decided that she wanted to document their experience by writing about her time in Cape Verde and about the people she met there.

Judy Aulette joined her soon after when she and Carter started to exchange ideas about how to conduct the research, what questions to ask, and how to understand and record the Cape Verdean voices. Work at this point, as we go to press, is so much a synthesis that it's difficult to say which author wrote which words. But the original idea and the supervision of the data collection as well as the ethnographic insights come from Carter. Aulette was central to the conceptualization of the issues as research questions and the inclusion of the political and economic context as essential to our understanding of Cape Verde.

The project has evolved through many stages. In the development of this book, we originally intended to write just about our study but as the venture grew we decided that it might be useful to other ethnographers, especially those just starting out, if we explicated the process by which we developed this research and this manuscript. We, therefore, decided to include at the end of each chapter a reflection on the process of the work.

The book is most importantly an expression of life in Cape Verde through the ideas of Cape Verdeans. But it is also a reflection by the authors on how to understand that expression from the point of view of two people from outside Cape Verde. In addition, it is a description of our journey developing a research question, gathering data, deliberating about how best to understand the information, and determining where all of this fits into our understanding of women, Cape Verde, globalization, oppression, and resistance. Our goal, however, was not to stop at the level of description and interpretation but to offer our ideas to the quest for change.

We hope that this book will introduce people to Cape Verde. We also hope that it will help them to see where Cape Verdean women fit into the global political economy. In addition we hope that students of qualitative methods and critical ethnography will gain insight into the techniques, difficulties, and effectiveness of these methodologies. And, most of all we hope that our findings can become part of the struggle to make globalization a tool for equality and justice rather than exploitation and oppression.

Katherine Carter is a sociology professor at the University of Kurdistan-Hawler in Northern Iraq, where she teaches Introduction to Sociology, Social Inequalities, and Introduction to Ethnography. Carter would like to thank Minichel Alemu for his continuous love and support during the long months of writing and to thank our baby boy, Abraham Isaac Alemu, born while this book was going into production. She would like to thank members of the Sociologists for Women in Society and Association of Black Sociologists who, in 2005, heard Carter's presentation on Cape Verde and gave important feedback. She expresses sincere gratitude to students and colleagues of the Ethnography Department at the University of Debrecen, Hungary, who also provided valuable insights. Finally, Carter expresses great appreciation to Judy Aulette, whom she met in 1997 in a graduate course at UNCCharlotte. Aulette was teaching Social Stratification and Carter was her student. She was also Carter's M.A. and Ph.D. advisor throughout the next ten years. Aulette provided constant encouragement and challenged Carter to take risks in writing and publishing, and in living and teaching abroad. Carter thanks the students, friends, and families of Cape Verde who made this book possible.

Aulette is a professor in the Department of Sociology and the Women's and Gender Studies Program at UNCCharlotte where she teaches and writes about family, gender, women, and activism. Aulette would like to thank Albert Aulette, Anna Aulette-Root, and Elizabeth Aulette-Root for their reading of several drafts of this work and their constant support; Tamera Shefer, director of the Women's and Gender Studies Programme at the University of the Western Cape in Cape Town South Africa for introducing her to gender in Africa; Laurie Garo and Christine Davis her colleagues and members of her writing circle at UNCCharlotte for giving her good advice about how to better communicate her ideas; and the members of the Sociology Department at University of the Western Cape who heard one of the first presentations on the work and told her what she needed to do to connect contemporary Cape Verde to its colonial history. She is also grateful to Luba Ostashevsky for agreeing to publish the book. And most of all she wishes to thank her dear friend Katherine Carter and the students and women who participated in this study as insiders and outsiders.

1

About This Book

Cape Verde, an island country 450 kilometers off the coast of Senegal, West Africa, has been swept into globalization just as every other nation in the twenty-first century. Through in-depth interviews, this research looks at what Cape Verdeans have to say about women's lives today, their economic and personal difficulties, and the ways that women do not accept their fate but actively engage to challenge their surroundings. Women form the core of Cape Verdean society—there were over 13,000 more women than men in Cape Verde in 2006. Awareness about women's needs and challenges is essential not only to understand their experience but also to understand Cape Verde as a whole. Furthermore, as we will see in this manuscript, understanding Cape Verde women may help us to better understand globalization and routes to transforming globalization into a force for improving the lives of people in the Global South.

Women in Cape Verde face many problems associated with globalization such as poverty, managing single mother–headed households, and violence. This book looks at women's frustrations and joys and explores the ways women resist as well as cope with the difficulties globalization has brought to them. We examine the more conventional ways women are organizing around the problems they face, but we focus, in particular, on the less visible and often unacknowledged forms of resistance they utilize: *batuku* dancing and Creole language. We conclude that scholars need to reconceptualize the concept of resistance in order to "see" the lives of women in nations such as Cape Verde and especially in order to identify the bridges to political change.

Economics in Africa

African nations are part of what we now refer to as the Global South (Eisenstein, 1998). Scholars have long debated how to conceptualize and what terms to use for blocks of nations in the world that differ economically and in regard to international power. Terms like first world, third world, developed, underdeveloped, and developing are part of this history. The United Nations, for example, identifies 128 nations in Africa, Latin America, and Asia as "developing." The "least developed" of these include countries such as Afghanistan, Cambodia, Niger, Rwanda, Haiti, and Yemen, which face extreme poverty and most of whose citizens lead difficult lives. Because most of the nations with large populations of poor people are in the Southern Hemisphere while the richest nations are largely in the north, the terms Global South and Global North have come to indicate the gap between countries. Today, most scholars use the word Global North and South to distinguish between nations in the northern hemisphere that are generally wealthier and more powerful and nations in the southern hemisphere that are poorer and less powerful.

Cape Verde is a nation of the Global South. Although it recently was removed from the list of "least developed nations," it remains poor with 36 percent of its population living below the poverty level and its continued reliance on grants and loans to maintain its economy (*Africa Research Bulletin*, 2007).

Along with problems of hunger, lack of access to water and health care, difficulties finding safe reproductive services, and limited opportunities for education, African nations, like many other nations in the Global South, endure severe economic problems and increasing dependence on other countries. The United Nations estimates that nearly half of the world (44%)—about 3 billion people—in the world live on less than $2 a day, and because of poverty, millions lack access to even basic human needs such as water, food, sanitation, and literacy (New Economic Foundation, 2006).

Table 1.1 shows the proportion of the population that lives on less than $2 a day in different regions in the world. The table shows the enormous numbers of people in poverty and the large diversity from one region to another in regard to the proportion of people who are poor. The table also shows that Sub-Saharan Africa, along with South Asia, is the very poorest area and the proportion of poor people in these regions, more than 3/4 of the total population, is striking.

Table 1.1 Percentage living below $2 a day poverty line in the world by region, 2001

Region	*Population (in Millions)*	*Proportion Living below Poverty Line (in %)*
East Asia and Pacific	1,823	46.4
Eastern Europe and Central Asia	474	19.1
Latin America and Caribbean	519	25.2
Middle East and North Africa	279	23.2
South Asia	1,378	77.7
Sub-Saharan Africa	674	76.2
High-Income Nations	982	0.0
World	6,128	44.3

Source: New Economic Foundation, 2006.

At the same time so many people are poor, a few people have accumulated great wealth. The assets of the 200 richest people in the world are greater than the combined wealth of the poorest 40 percent of the world. That gap continues to grow as the wealthiest 5 percent of people in the world receive 82 percent of the income (Munck, 2005; Randerson, 2006).

Gender also plays a role in people's economic troubles. Women in the Global South, like women in the Global North, are often lower in status, power, and access to resources than are men. For example, women provide 80 percent of the agricultural labor in nations such as Uganda and most of the rest of Africa but they are much less likely to own land. Only 3 percent of the women in Uganda own land and this is not unusual in the Global South. If women do not own land or other property like cattle, they cannot participate in development programs or receive loans and credit. Only about 1 percent of loans to develop the land are made to women in Uganda, for example (Wamboka, 2002).

Women work hard but they are even more likely than men in poor countries to live in absolute poverty. Per capita income levels in Africa have decreased since 1973 and trade has declined. Since the 1980s, furthermore, the proportion of men living in absolute poverty has increased by 30 percent while the proportion of women in absolute poverty rose 50 percent (Emasu, 2002).

Global Debt Crisis in Africa

These problems of poverty and the growing gap between the rich and poor have been exacerbated by international economic pressures. One of the most important of these in the Global South is debt. Cape Verde is a debtor nation. It owed $360 million in external debt in 2002 (the most recent year available) and pays about $14 million a year in debt service (CountryWatch, 2007). The global debt crisis continues to grow, especially in poorer nations. In 1974 the total debt in the world was $14.8 billion, by 1989 it was $143.2 billion and in 2004, $300 billion. Part of this increase in debt is a result of additional loans, but much of it is because of growth in the interest owed.

Between 1970 and 2002, for example, the continent of Africa received $540 billion in loans from wealthier nations and international financial institutions such as the World Bank. The debtor nations have paid back $550 billion of their debt but they still owe $295 billion. Yes these numbers are correct. The difference, of course, is a result of interest. When borrowers take a loan they agree to pay back the amount they borrowed plus the interest accrued. The relationship between debtor nations and wealthy nations or the international financial institutions wealthy nations control creates a situation in which more money is transferred from poor countries to rich countries than vice versa (Argus, 2004; MacDonald & Gibson, 2007; Makwana, 2006; Parks, 2006).

For all but three of the past 23 years, countries in the Global South have paid out more money in the form of interest, repayments, penalties, and fines on old debt than they have received in the form of new loans. Although almost all poor countries have repaid more than they borrowed, their debts continue to mount. The increasing debt and debt payments cause poor nations to divert resources away from spending on critical human needs such as health, education, and food with dire results (Kapoor, 2006). For example, UNICEF angrily asserts, "hundreds of thousands of the developing world's children have given their lives to pay their countries' debts, and many millions more are still paying the interest with their malnourished bodies" (UNICEF, 1990: 30).

In addition, economic relationships among nations tend to benefit the wealthier countries (Kerr & Sweetman, 2003). Some of the land in Cape Verde, for example, could be planted and harvested to feed people but is planted instead with cash crops for foreign exchange. Due to indebtedness and foreign dependence, fresh produce is regularly sold or changed to a nonperishable type such as canned tuna for export rather than consumed by the population. Widespread malnutrition is one of the effects of this foreign dependency (Ludtke, 1989; Bigman, 1993).

Sen and Grown (1987) call this denationalizing of the Global South. Foreign trade and business get in the way of the freedom of local governments. Many of these economies become dependent on single-product export trade, making them weak and vulnerable to market changes. Such dependence, as opposed to self-sufficient development programs, has more often than not been pursued. Instead of development, political oppression, hunger, and weak export-focused economies have resulted (Acosta-Belén & Bose, 1995).

The effects of the growing burden of debt and the poverty associated with it are life-threatening in all of Africa and the crisis is escalating. Over 50 percent of Africans continue to live on less than $1 per day (*National Geographic*, 2005). In 1990, the World Bank predicted quite accurately that poverty would decrease in every third world region in the year 2000, except Africa, where poverty would increase (Dixon, 2004; Callaghy, 1993). And this trend continues into the twenty-first century as some developing countries are predicted to reduce their poverty rate by 2015, but the forecast for most of Africa is that it will fall farther and farther behind (Civil G8, 2006).

Globalization

The whole world, including Cape Verde, has been shaped by these forces of globalization. But what exactly is globalization? Chandra Mohanty (2003) writes, globalization refers "to the process of corporate global economic, ideological, and cultural reorganization across the borders of nation states" (p. 272).

This definition sounds fairly benign, but huge debates center on the meaning globalization has for different nation states and the people who live within their borders. Organizations such as the World Trade Organization (WTO), the International Monetary Fund (IMF), and the World Bank have heralded globalization as a great movement that will remedy problems. They argue that globalization will benefit everyone as it spreads the economic and social systems of the wealthiest and most powerful nations all over the world (Callaghy, 1993; Held & McGrew, 2000; Petras & Veltmeyer, 2001).

Alan Greenspan (2000), who was chairman of the Federal Reserve Board in the United States from 1987 to 2006, was famous for his convoluted manner of speaking. In a speech at an international gathering of financial leaders in Mexico City, however, he was surprisingly straightforward when he flatly stated, "Although globalization has its critics, I say with some conviction that the increasing interaction among national economies has engendered benefits that have significantly exceeded their costs over the years."

The CATO Institute (2008), a conservative political think tank in the United States further explains on its website the benefits of globalization from its point of view:

> Globalization describes the ongoing global trend toward the freer flow of trade and investment across borders and the resulting integration of the international economy. Because it expands economic freedom and spurs competition, globalization raises the productivity and living standards of people in countries that open themselves to the global marketplace.
>
> For less developed countries, globalization offers access to foreign capital, global export markets, and advanced technology while breaking the monopoly of inefficient and protected domestic producers. Faster growth, in turn, promotes poverty reduction, democratization, and higher labor and environmental standards. (CATO 2008)

This sounds like good news for the world's people. Does everyone agree with the Federal Reserve Board and the CATO Institute point of view? Is globalization the answer to the world's problems? Do the multinational corporations which are the hallmark of globalization help or hurt developing countries? How can globalization be the answer if poor nations are becoming even poorer and nearly half (44%) the world's people are living on less than $2 a day?

According to Anup Shah (2001), a critic of globalization, globalization is not the answer; it is in fact the problem. He argues that the sophisticated legal and financial strategies that overpower the boundaries of local laws and standards in developing countries of the South serve only to pull the labor, products, and services into the hands of wealthy people and corporations largely in developed countries of the North (Shah, 2001).

Other critics of globalization (Kerr & Sweetman, 2003; Rai, 2002; Ault & Sandberg, 2000) assert that the benefits of globalization are only for those at the top who live in the wealthy nations of the world while those who live in poorer nations see nothing but economic hardship and a destruction of indigenous culture and environment. They further argue that globalization, controlled by institutions such as the IMF and its Structural Adjustment Programs (SAPs) and Poverty Reduction Strategy Papers (PRSPs), makes life more and more difficult for poor people, especially for women (Cagatay, 2003). They maintain that colonialism in the Global South continues today under these programs. Decisions are made that cause economic problems for nations of the South while benefiting nations in the North.

Women, because they are at the bottom of their communities are the most likely to suffer (Acosta-Belén & Bose, 1995; Rai, 2002; Ault & Sandberg, 2000).

The World Bank and International Monetary Fund (IMF)

- In 2003, Cape Verde accepted a loan of $11.5 million from the World Bank.
- In 2007, the country received an additional $3.0 million (World Bank, 2007).
- In 2005, Cape Verde received a loan of $110 million, distributed over five years, from the Millennium Challenge Corporation (Millennium Challenge Corporation, 2008).
- In July 2008, Cape Verde joined the World Trade Organization (AGIPNEWS, 2008).

Almost 125 million dollars and a place in the most powerful international economic organization sounds like a good deal for nations such as Cape Verde that are struggling to develop their economy by building infrastructure and businesses and strengthening their labor force. The problem is that along with international aid and assistance from these financial institutions come strict conditions for economic reform.

The World Bank claims that its goal, with the help of the IMF, is to aid nations of the Global South in their efforts to reform their economies so that they can develop and become equal partners in the global economy (World Bank Development Report, 2004). In order to achieve these goals, the IMF has called for reforms such as facilitating increases of foreign investment, opening markets and trade, and building export-centered economies (Ault & Sandberg, 2000). Nations that do not abide by the recommendations and conditions of the IMF will find themselves unable to secure loans from the international community. In addition, nations which are not in compliance with IMF conditions generally cannot get private funding from international banks and most countries need this just to be able to refinance at the end of the loan period.

The IMF exists in over 185 countries. Money for loans to poor countries originates from deposits made by the member countries of the IMF. The weight of member countries' votes within the IMF depends on the amount of their financial donations. The strict conditions each country must follow in order to obtain a loan, therefore, depend on the approval of the countries that make the greatest deposits.

The United States has the largest voting share (17%) as it contributes about $39 billion. France (5%), Germany (6%), Japan (6%), and Britain (5%) also make large contributions and together these five nations control nearly 40 percent of the vote (International Monetary Fund, 2008). As a result, they carry great authority over the conditions attached to each loan awarded to any of the other 185 countries, along with its policies (Ault & Sandberg, 2000; International Monetary Fund, 2007). The decisions about what should be done in the nations applying for a loan—usually the poorest nations—therefore, is decided by the political and economic leaders in the wealthiest nations, who are heads of major corporations or obliged to those corporate leaders for campaign funds that financed their seats in government (Petras & Veltmeyer, 2001). In sum, power is concentrated at the top in the hands of the wealthiest people in the wealthiest nations who make life and death decisions about the fates of the poorest people in the poorest nations.

Structural Adjustment Programs (SAPs)

Structural Adjustment Programs (SAPs) are common types of conditions the IMF demands from debtor nations. SAPs require nations to reduce public spending on social services and to privatize many goods or services currently controlled by the government. According to the World Bank, SAPs are supposed to aid poor nations by helping them to set goals, create priorities, and make their economic decisions and actions more focused and efficient. The reforms called for by the World Bank and IMF and SAPs, however, have proven devastating for countries in the Global South and especially for poor women living within those nations (Ault & Sandberg, 2000; Rai, 2002). Poor women have taken the hardest blow from these programs because they most often depend on the social services and must make do when governmental provisions of health, education, and food are taken away.

Research in Zambia, for example, finds that women have suffered greatly as a result of SAPs (Ault & Sandberg, 2000). Changes required by SAPs have led to extensive unemployment; considerable price increases; suspension of education for many, especially girls; a dramatic decrease in access to health care; and increased social class stratification. Between 1980 and 2000 primary school enrollment rate fell from 96 to 77 percent in Zambia. Half a million children are now out of school, out of a total national population of only 9 million. And less is being spent on each pupil's education. In 1991, the Zambian government spent about $60 per primary school pupil; it now spends just $15 (Lynsap, 2000). Girls are especially likely to be pulled out of school as parents can afford to pay tuition for only some of their children (Jere-Mwiindilila, 1994).

While these problems with SAPs affect nearly all citizens regardless of gender, they make life especially difficult for women as they have the burden of work both inside and outside the home, making a living and raising children, and in addition seeking extra resources when social service provisions are stripped away (Rai, 2002). Research detailing the affect of SAPs implemented in Cape Verde has not yet been conducted. Nevertheless, Cape Verde has many similarities to Zambia and we can assume comparable problems developing there as a result of SAPs, especially for women.

In 2005, the Foreign Ministry of Cape Verde issued an official statement regarding its view of the external debt and the restrictions it places on economic progress in Cape Verde. Ministry officials wrote,

> The impact of the [government's development] strategy, particularly on living conditions of the poorer population, would be far greater if the country did not have to spend a considerable amount of resources on external debt service. Alternatively, these resources could be put to much better use in priority sectors such as Education, Health and Basic Infrastructure...the Government calls on the international community to relieve Cape Verde of the heavy debt burden by means of its cancellation.... More than a reward for good behaviour, what is called for here is that a country not be penalised for its good performance in the area of development. (Official statement 2005, supplied by Foreign Ministry of Cape Verde, quoted in Baker, 2006)

Still Sapping the Poor: PRSPs

In the late 1990s, the term SAP was changed to PRSP which stands for Poverty Reduction Strategy Papers and new features were added to increase control of the adjustments by local governments. The World Bank asserts that increasing local government's participation in creating the policy will lead to greater commitment to the loan programs, thus better fiscal policy. The content of PRSPs, however, has turned out to be similar to the original content of bank authored SAPs (Stewart & Wang, 2004).

Anup Shah (2007) writes PRSPs in fact do not focus on poverty reduction and do not involve local government or civil society in discussions on economic reform. Jubilee research (formerly Jubilee 2000) in addition maintains that the previous neoliberal, economic policies have not been changed in the transition from SAPs to PRSPs. In 2000 when explaining PRSPs to a group of finance ministers representing various countries in Africa, the IMF demanded that their governments privatize faster and organize trade liberalization more quickly—conditions that even Joseph Stiglitz criticized when he was head of the World Bank. Charles Abugre (2001), director of the Integrated Social Development Center in Ghana writes:

> PRSPs are a classic case of empty rhetoric. It could provide a useful veil for the World Bank and IMF to continue their neo-liberal agenda unseen. Judging by the gaps between what the IMF and World Bank say, and what they do, PRSPs may well result in the worst of both worlds for poor economies and poor people within them, by legitimizing and institutionalizing yet additional conditionalities without significant benefits either by way of debt reduction or real change in the content and "ownership" of policies. Indeed, PRSPs and the preoccupation with them have already provided these institutions with the excuse they need to simultaneously delay action on significant debt relief and deflect attention away from their structural adjustment programs.

Globalization has tied the world together in a way that perhaps has the potential for ending poverty and promoting equality and justice but thus far it appears to have primarily created problems. Poverty and misery abound and the policies of loans, SAPs, and PRSPs exacerbate the problems.

The problems associated with globalization and the colonial history out of which globalization emerged have not gone unnoticed by African women. They have surely suffered greatly from the poverty, hunger, and cuts in education, welfare, and other public services in housing, water, and sanitation. They have also responded not only by surviving in spite of these problems but by organizing and fighting back.

Waves of African Feminism

In the West it is common to think about women's activism as fitting into three waves. In the United States, feminism has been seen as divided into a first wave that began in the nineteenth century and primarily focused on suffrage. A second stage is the Women's Liberation Movement of the 1960s and 1970s. The third stage is the feminist activism organized around the diversity of women especially by race ethnicity that emerged in the last quarter of the twentieth century. The third wave includes all of the branches and fractions that split off from the Women's Liberation Movement which was largely organized around white middle-class women's lives.

African women's activism has also been divided into waves, but the turning points and the focal issues are quite different from the three waves identified with the American women's movement. Using the model of the waves, scholars of African history have proposed a postcolonial timeline of African women's activism running from 1950 to 1970. During this period, the focus was on development issues such as "integrating women into development, promoting women's cooperatives, developing small-scale industries, training rural women to take on leadership positions, and establishing national mechanisms for integrating women in development projects" (Adams, 2006, p. 189).

A second wave of regional activism in Africa began in 1980 and continues today. This period is marked by global debt, structural adjustment, and economic crisis. The globalized economy introduced problems to the people of Africa, but it also facilitated international communication and organization that helped African women, for example, by increasing the growth of non-governmental organizations (NGOs) committed to solving women's problems and providing alternatives to the official views and activities of governments. Peace and the elimination of poverty have dominated the agendas of these NGOs.

At the beginning of the twenty-first century, another turning point emerged for African feminists suggesting that perhaps a third wave is developing. In 2002, the launching of the African Union (AU) gave impetus to the struggle for gender justice. The AU is modeled after the European Union and "seeks to promote unity among African countries and peoples, political and economic integration, peace and respect for democracy and human rights" (Adams, 2006, p. 194). A centerpiece of the AU is its commitment to gender equity. Women make up fully half of the executive body of the organization. In 2003, the AU passed the Protocol on the Rights of Women in Africa, which explicitly endorsed laws and policies that create quotas to increase women's representation in decision-making bodies, prohibit female genital cutting, set the minimum age of marriage at 18, and guarantee a woman's right to terminate a pregnancy in the case of rape, incest, or to save the mother's life (Adams, 2006).

Cape Verdean Women Respond

This book investigates the effect of the massive worldwide forces on women in Cape Verde, a group that is outside the central control of the world, of much of the nation, and even of their own households during the transition from the second wave to the third wave of African women's history. The research provides insight into Cape Verdean women's lives by placing their lives into the larger global context, questioning how globalization has shaped women's personal stories, and exploring the ways women are resisting these challenges.

Cape Verdean women have endured severe economic and personal difficulties due to the economic context that creates job shortages and low wages. In addition to the direct effect of these economic difficulties on women, there have also been indirect effects. The most important of these is out-migration especially of men. Recently, migration patterns have changed and women are joining men in their search for jobs outside Cape Verde. Men, however, have been and continue to be the bulk of emigrants (Carling, 2004b; 2002a).

The desire to work abroad is common in Cape Verde. Quarterly employment surveys conducted by the government ask about the wish to emigrate and typically more than half of the respondents say they would like to emigrate. Those who respond are usually younger and have lower levels of education. They mention unemployment, having relatives abroad, and receiving remittances as the key factors in their desire to relocate (Carling, 2002a). Lower unemployment rates and relatively high wages, especially in the United States and the European Union, are a constant draw for people all over the world seeking economic opportunities and even survival. The largest numbers of Cape Verdean émigrés are living in the United States, Portugal, France, Netherlands, and Italy.

Women, in Cape Verde, who experience economic difficulty and the challenges of single-parenting, respond to their problems through standard methods of running for office, maintaining political organizations and educating each other about political issues such as violence against women. In spite of women's best efforts, however, these activities remain fairly small and quiet for a number of reasons including the dominant ideology in Cape Verde that tells people dealing with violence that their problems should be settled in the home, not by the state. In addition, material barriers such as transportation and funds to file reports pose problems for women. Our project explores the conventional organizing efforts of Cape Verdean women against violence and these are described in chapter 6. We highlight, however, the more widespread but less conventional ways women are challenging the status quo through culture as revealed in chapters 7 and 8.

Our research examines the role that *batuku* dance and Creole proverbs play in Cape Verdean women's resistance to the harmful aspects of globalization. Perhaps you had not thought of these kinds of activities as resistance. The meaning of these activities to the women as insiders, however, may be very different than the meaning of these activities to outsiders. From the point of view of Cape Verdeans, through *batuku* song and dance and spoken Creole, women are able to acknowledge their difficulties, develop alternative views of their world, and change their personal circumstances. They develop a deeper sense of self that contrasts with the dominant, patriarchal, and neocolonial discourse. This resistance takes the form of everyday acts of rebellion which includes language and dance (Heath, 1994).

The Structure of the Book

We came to this project asking what is life like for women in a nation of the Global South? What problems do they face and how do they cope with or challenge that experience? In this book we begin by laying out the social context of life in Cape Verde and then explore the ways women are responding. We conclude by reflecting on the significance of their response to our understanding and our activism.

Chapter 2 provides a context for understanding life in Cape Verde by looking at its history and some of the prominent features of its culture and economy today. The chapter describes Cape Verdean history, geography, religion, and industry. Chapter 3 explains how we went about studying the island nation; including the methods used to gather information, and the problems we faced collecting data. We used a complex methodology involving layers of participants and we used a combination of tools for analysis including critical ethnography and critical discourse analysis. Chapter 4 develops many of the themes from chapter 1 by more closely examining the specifics of the influence of globalization on Cape Verde. Our focus in particular is on the way in which globalization has shaped economic issues in Cape Verde and the role of the government in addressing those issues. Chapter 4 presents empirical data from interviews with people of Cape Verde as well as survey data collected from the general population. Chapter 5 examines the situation of Cape Verdean women within the economic context described in chapter 4. In our interviews with Cape Verdean women, we ask what their perceptions of the situation are and how the economic context of globalization is affecting their personal lives at work and in intimate relationships. Chapter 6 explores another important feature of women's lives in Cape Verde today: the problem of violence against women. Through a review of international, regional and national policy as well as open-ended written surveys of women activists, we investigate women's experience and their efforts to combat violence by way of political action. Chapter 7 introduces the less conventional ways women respond to and resist the problems they face. Here we uncover *batuku* song and dance as a vehicle of resistance for women. We look at the ways women express themselves, cope with, and challenge the difficulties in their lives through dance. Chapter 8 examines the second less conventional mode of resistance we discovered—language. Through a critical discourse analysis, we dissect the Creole proverbs and sayings that women use to challenge the power relations in society. Chapter 9 concludes with a discussion of the theoretical and political implications of the pervasive defiant activities of Cape Verdean women with a focus on the concept of resistance and the need to find connections between apathy and despair and effective organized movements for social change.

Our book presents a study of women in Cape Verde—the problems they face and the creative ways they address their problems. The book also tells a parallel story of our efforts to explore these issues. At the end of each chapter, the reader will find a section called Reflection on Research Methods. Each of these reflection sections describes and considers the tools and techniques, steps and procedures that we followed to conduct our research and prepare this book. How did we enter the community, gather information, analyze data, and write up our results?

Reflections on Chapter 1 Research Methods

Nearly every report on social science research follows a format which begins with an introduction, moves through a review of the literature of previous research, describes the way the data were collected, presents an analysis of those data, and finally draws conclusions about the significance and meaning of the findings. This book follows this format beginning in chapter 1 by introducing the topic and presenting arguments about why this is an important topic to study and an interesting topic for the reader to consider. The introduction provides a little background on the general issues and it presents the specific research question to be addressed in the study. Our research question is two-fold: what problems do Cape Verdean women face in the globalized world? And how are they responding to those challenges? You can see in this introduction that we begin with the broad issue of economic problems associated with globalization in the continent of Africa and move to the specific question of how those problems are being played out in Cape Verde, especially in Cape Verdean women's lives. The second aspect of the research question, the issue of how are women responding to the problems they face, is then presented.

This first chapter also provides a structural introduction telling the reader how the rest of the book is organized by giving an overview of the book and providing brief summaries of the chapters. In this introduction we have tried to capture the attention of the readers and provided them with a hint about what they will explore as they read through the rest of the book.

2

Snapshots of History and Life in Cape Verde

The streets of Cape Verde are filled with people talking, working, and playing. Children shoot marbles and wind strings around their hands, women swap gossip and news, and men deal cards and drink beer. Even a quick glance at Cape Verde, however, shows women at the center of the picture. Everyday women are outdoors cleaning fish and washing clothes by hand. Other women carry baskets of fruit, vegetables, or fish on their head for sale, shouting as they walk, *"atun freska!"* (fresh tuna). Two women take turns crushing and pounding corn with a pestle and mortar. Students in uniforms rush off to class, attempting to catch the local bus already packed with riders. The call of the bread woman causes a crowd to gather round in the cobblestone streets. Women walk back from the *chafaris* (community water channel), slowly steadying the plastic jug on their heads, which each hold 25 liters of water.

Women fill the streets. But the women who are working and socializing outside in the neighborhoods and markets are nearly all working-class women while one group of women is largely invisible. Wealthy women are not often seen in the streets. Their maids handle public errands, grocery shopping, taking care of children outdoors. If they travel they go in cars, rather than walking or public transit and their social excursions are behind closed doors in large hotels or exclusive restaurants and clubs. Wealthy people comprise only a small percentage of the total population although as we shall see in later chapters on the political economy of Cape Verde, they play important roles in the organization and experience of globalization.

Greetings between people follow a protocol. One commits a small crime if she or he accidentally neglects to greet someone properly. Special inquiries or visits are made pertaining to those suffering illnesses. People exchange conversation about the weather or family news. Breaking off a conversation is equally ceremonious, aiming to evade abruptness. The leave-taking person announces they are going multiple times before actually departing. The other person replies, "oh, it's early, don't go yet." The relaxed concept of time, appointments, and delays is characteristic of life on these islands. Appointments between friends are not common. Social encounters appear to just happen naturally; they do not seem to be planned although the assumption is that their paths will cross frequently. If two acquaintances do set a date, and one does not show, nothing is said the next time they see one another. No social norm has been broken. And there are no hard feelings.

The idea of *fazi vontade*, free will, or one does as they wish is primary. Generally, Cape Verdeans do not live their life by the clock but by their own personal desires and feelings. One will not show up for appointments, or carry through on a commitment if one does not have the true desire to do so. An outsider might ask, "Well, how does anything get done? If we all lived like that, then nothing would be achieved or accomplished."

The pieces fall into place easily when both parties show up to carry out the task, whatever it might be, socially or professionally. Both know that the other wants to be there fully and completely, with genuine intentions. When questioned about this cultural practice, a Cape Verdean said, "How many times have you attended a meeting thinking, 'Why did I come?' 'I'd rather be any place than here.' " "Our behavior reflects our thoughts," she explained, "and in this country one does well what he or she believes and feels." Obligation and commitment are not artificial or forced. They are either heartfelt or they are not abided by.

Gossiping between neighbors and especially talk of who has "*arranjado*" with whom—who is dating whom—is common. Young men and women more often have a boyfriend/girlfriend than not. Being single or alone is viewed negatively. Young men approach women at a social gathering such as a party and engage them in conversation or ask them to dance. Sometimes the man will visit the woman or attempt to see her again. But most times, he will show patience and wait for the time that their paths cross again before pursuing her. Being "just friends" between young men and women is not common and is not understood. Either a couple is *arranjado* or they have no relations at all.

Style of dress varies throughout the country and especially by age. Elderly women dress more conservatively and wear head scarves and skirts or dresses. Young girls and women wear short skirts, tank tops, and stylish, fitted shirts. Young men wear trendy clothes, sometimes sent from abroad.

When Cape Verdeans speak about their country they express feelings of approval. They take great pride in their land, their hospitality, the climate of the country, their lack of involvement in wars, and their beautiful beaches and mountains (Solomon, 1992). When meeting a new person, they share their joy about the country. The motives for people from other nations living and staying in Cape Verde are not questioned. In the eyes of a Cape Verdean, of course anyone would want to live in their unspoiled and peaceful country.

People have a strong sense of community spirit. Despite water shortages and lack of access to education, opportunity, and jobs, Cape Verdeans have a culture of prevailing grace, dignity, and open

Figure 2.1 These women receive Christmas presents during a USAID sponsored event in Praia, Santiago. Most of the elderly women cover their heads in public, while young girls show off a modern look. The woman dressed in black is in mourning.

and warm human relations (Kutsche & Ness, 1981; Solomon, 1992). They work hard to earn a living and to support their families but they also make time to share a joke, speak with proverbs of insight, create wild music, dance until dawn, and write touching poetry (Solomon, 1992). They frequently make the best of their situation, drawing on every moment of tranquility and contentment. Small celebrations are grounds for throwing a *festa* (party) and to dance (Solomon, 1992). The fact that Cape Verdeans "play hard" may not only be because they are culturally celebratory but also as a response to injustices—they need something to take them away from their awareness (because of globalization) of the lack of opportunity (even if they don't say this out loud). And as we shall see later in the book, these celebrations may themselves be expressions of resistance to the difficulties they face.

Celebrations and parties are important, whether it is a birthday, baptism, graduation, arrival, departure, holiday, engagement, birth of a child, or just because. The year is full of religious and national holidays, which lead to big celebrations full of special food and drink and entertaining friends, family, and visitors coming from abroad, especially for Christmas and New Years. After New Years, the season of festivals begins, which are generally Catholic saints' days. Religious and nonreligious people participate in the occasions, which have both African and European inspired customs (Solomon, 1992). They begin with church services and include processions, drumming, and eating specially prepared food. Traditional Cape Verdean dishes are especially eaten on Ash Wednesday. Special foods are grown or made throughout the country: kale, sauce made from coconuts, potatoes, dishes made from corn, then added to a plate of fish. It is a special day if one wants to taste traditional food.

Geography

Cape Verde can be called a country of contrasts. It is cosmopolitan but it also is isolated far from the nearest other nation, Senegal, from which it gets its name (Irwin & Wilson, 1998; Solomon, 1992). Cape Verde was given its name not because of the greenness of the islands themselves but after the green end of Cap Verd, Senegal, the most western point of Africa. Cape Verde's 10 islands are divided into two groups: Barlavento or Windward Islands in the north are Santo Antão, São Vicente, Santa Luzia, São Nicolau, Sal, and Boavista. The Sotavento or Leeward Islands, to the south, are Maio, Santiago, Fogo,

Figure 2.2 This volcano on Fogo Island erupted 29 times between 1563 and 1995 (Almeida, 1997a).

and Brava. There is a wide contrast of elevation in the country: Fogo's volcano reaches 2,829 meters while Sal Island rises only 406 meters.

The islands are composed of mountains, beaches, and sand dunes (Irwin & Wilson, 1998). They are beautiful and picturesque but they are not fertile and they do not contain significant amounts of minerals to be mined. The terrain is steep, rough, and rocky. Natural resources include salt, basalt rock, limestone, clay, and fish. Industries process fish and fish products, clothing, construction, building materials, and ship repair. Agriculture products consist of coffee, corn, beans, bananas, and especially sugarcane.

The average rainfall per year in Praia, the capital, is 24 centimeters (9.5 inches). The climate is moderate with a warm and dry summer (Nationmaster, 2000). Vegetation is sparse. Town or city dwellers do not have gardens in their homes although the rural populations are subsistence farmers who own small plots of land and grow food on them for home consumption. In 1988, 32,193 individuals owned land, which included 20,539 men and 11,654 women. Though men own more land than women, women outnumber men working the

Figure 2.3 A man cuts and weighs bananas on the island of Santo Antão. Bananas are a main agricultural export product throughout the islands, but they also cause enormous destruction to the soil.

land. In 1988, 89,229 individuals worked the land, which included 41,862 men and 47,367 women (Cape Verde Agricultural Census, 1988).

History

The isolated location of Cape Verde has been a feature of the tragic aspects of its history. Other nations used and abused the islands for many centuries. History shows Cape Verde a flourishing and successful place when it was useful to other countries and an unsuccessful and disastrous place when those other countries decided they didn't need Cape Verde any longer and abandoned it (Irwin & Wilson, 1998). Furthermore, it shows that those moments of "success" were largely a result of the slave trade.

Although sailors from Africa stopped by the islands for centuries, until the colonial period, Cape Verde did not have permanent human inhabitants. In 1460, Diogo Gomes, a Portuguese explorer claimed Cape Verde as a colony for Portugal and it remained under Portuguese rule until 1975. In the 1500s, other Europeans as well as the Portuguese began to use the islands as a place for ship repair and supplies, chiefly water and food. Another group of people were also among the first to settle Cape Verde, Jews from Europe. During the thirteenth and fourteenth centuries, Jews who had been driven out of European nations, such as Portugal during the inquisition settled in Cape Verde.

Slaves from West Africa were a third group to become the first residents of Cape Verde. Nearly all contemporary Cape Verdeans trace their lineage to West African slaves who were brought to the islands and the Portuguese exiles who established and maintained the slave trade. During the centuries of the slave trade, the majority of people who lived in the islands were slaves. A census, for example, in the late sixteenth century notes that 1,200 people lived in Praia: 1,000 were slaves, a few were free blacks, and the rest were white. In the interior plantations, 5,000 were slaves and 1,000 were listed as residents, meaning they were not slaves (Curtin, 1969). Even the Catholic Church owned slaves to cultivate the land and maintain church buildings (Lobban, 1998).

Cape Verde played a major role in the slave trade because of its unique position midway between Europe, Africa, and the Americas and facing the Slave Coast. It became one of the crossroads in Transatlantic navigation and the slave trade and served as a center for slave trading. Thousands of West African slaves were kept on the island to work the land on sugar plantations and in cotton production and processing. Slaves, who had been professional weavers in their communities before they were captured and taken to Cape Verde, wove cotton into beautiful cloths, which were sent to West Africa and Brazil in the exchange for goods and people that were part of the slave trade between the Western and Eastern hemispheres. From the sixteenth century onward, with the significant expansion in the slave trade, Cape Verde became a transit point for slaves from several regions on the western coast of Africa (UNESCO, 2004). In the sixteenth and seventeenth centuries, more than 1,000 slaves per year were transported from Cape Verde. That adds up to more than 200,000 people total.

Both cotton and slaves became a major draw to the islands. Ships were obliged to stop at Cape Verde and buy cloth in order to trade in other countries. Cape Verde then functioned as a warehouse for slaves going to North and South America. Located in between Europe,

Africa, and the Americas, Cape Verde became an economically useful site with Portugal reaping the benefits. The demand for slaves went on throughout the 1600s and 1700s until Portugal was pulled into the War of Succession with France and Spain and the slave trade with the Spanish Indies ended (Irwin & Wilson, 1998).

As the demand for slaves decreased and later ended, Portugal no longer wanted to invest in the islands, which it felt could no longer give it anything in return. Cape Verdeans were left to survive day to day, without resources and little connection to the outside world, unable to plan or think about the future. Isolation and economic decline, furthermore, caused other major problems. Goats began to eat through the delicate vegetation. With no long-term management of the land, famine plagued Cape Verde. In every century there were one or two more famines than the previous century:

- From 1773 to 1776, 22,666 people or 44 percent of the population died from famine. Some people were sold into slavery in exchange for provisions.
- In 1830, 30,000 people or 42 percent of the total population died from famine.
- From 1854 to 1856, 25 percent of the population died.
- From 1900 to 1903, 11,000 people or 15 percent of the population perished of hunger.
- From 1941 to 1943, Fogo Island lost 7,500 people (31% of its population) and São Nicolau lost 28 percent of its population.
- From 1946 to 1948 Santiago Island lost 65 percent of its population to famine (Almeida, 1997a).
- From the mid-1990s, droughts cut the islands' grain crop by 80 percent, and in 2002 the government appealed for international food aid after the harvest failed (BBC News, 2008).

In addition to the suffering and death caused by lack of food, many victims of famine were tied to a system of debt peonage in the nineteenth century. Those who were unable to feed themselves were press-ganged into projects such as road construction in Cape Verde or were sent to other Portuguese colonies in São Tomé and Angola to work on cocoa and coffee plantations. Although slavery had officially ended, "vagrancy" laws were passed in 1875 that allowed the colonial government to round up citizens who were not employed and send them to these plantations to work. Vagrants were identified as anyone over seven years old. The laws perpetuated slavery well into the twentieth century (Ishemo, 1995).

A traditional Cape Verdean song illustrates both the suffering caused by the government's establishment of the debt peonage system. It also illustrates how culture is used to record that history and to protest the inhumane and unjust treatment, which is an important idea we will develop further in this book.

> Four o'clock in the dawning
> São Vicente folk are there
> To cry to their sorrow
> For sons who are sent away
> To São Tomé. (lament quoted in Ishemo, 1995)

In addition to famines and bondage, a lack of military protection led to continual raids and lootings of the islands by the English and French. Finally, when the steam ship replaced the sailing ship, Cape Verde's remaining function as fuel resupplier to ships came to an end, with only a few continuing to stop and reload with coal on the island of São Vicente (Irwin & Wilson, 1998). The raiding stopped but so did the trade with incoming ships when the shipping business fell in the 1900s. Soon Cape Verde's function as a coaling station disappeared completely as oil replaced coal as the energy on the seas. Few ships needed to stop in Cape Verde and it was not necessary to sustain big stations at the halfway point between Europe and South America. The world let Cape Verde go (Irwin & Wilson, 1998).

The collapse of the industry along with drought brought continual famine. Starvation forced Cape Verdeans to leave and many flocked to North America, especially northeast United States. Others found their way to São Tomé and Príncipe: 24,000 worked there between 1902 and 1922 and 34,000 from 1950 to 1970.

Catholicism

Christian religion has played a contradictory role in Cape Verdean history. It is a reflection of colonial control of Cape Verdeans through the promotion of European ideas and practices of religion, especially Catholicism since Portugal was a Catholic country. Religion, however, has also served as a source of hope and strength throughout the centuries as individuals battled against famine, drought, and migration. Catholicism was the dominant religion in Cape Verde from the landing of the Europeans in 1463 until the 1900s. Indigenous religions may have survived but records of them did not. Municipal authorities counted all citizens in the community as members of the Catholic Church. Christian ceremonies, holy days, and processions persisted as a central feature of life's everyday routines and struggles. The church kept the only official birth, marriage, and death records (Almeida, 1997b; Werlin, 1995).

In 1495, Our Lady of the Holy Rosary Church was founded in Cidade Velha, Santiago Island, the first permanent place of Christian worship in Sub-Saharan Africa. Later, a seminary, convent, and finally the first cathedral in Africa were built. Today, the United Nations has designated the cathedral as a World Cultural Heritage Site (Almeida, 1997b).

Catholic missionary activity in the islands began with the arrival of the Franciscan fathers in 1466 and Jesuits priests in 1604. The Holy Ghost Order arrived in 1941 and the Capuchins in 1946, followed by the Salesian fathers. These groups did not stay long, and did not have successful plans or missions. Church authorities in Europe were typically unhurried when having to replace priests in the islands. Priests and missionaries were lovingly accepted by communities, especially

Figure 2.4 A Church on Fogo Island. The calendar year is full of Catholic saints' days, marked with great celebrations, processions, and special foods.

in the rural areas. Some of the church leaders, however, were known more for the children they had fathered among the islands, rather than for their religious faith and passion (Almeida, 1997b).

In spite of missionary activities, Cape Verde was generally overlooked by the Vatican for almost three hundred years. At the same time, church leaders on the islands always came from abroad. Not until 1975 was a Cape Verdean allowed to lead when Dom Paulion Evora became the first Cape Verdean bishop. Today Catholicism is still strong. There are 48 priests throughout the islands and Pope John Paul II made a visit to the islands in 1990, acknowledging 500 years of Catholicism in the country (Almeida, 1997b; Nationmaster, 2000).

Thousands of Cape Verdeans migrated to the United States in the 1900s and joined Catholic Churches there. By 1904, many immigrants joined St. John the Baptist Church in New Bedford, Massachusetts, the first church established by Portuguese in the United States. In 1905, the community in New Bedford started their own church, Our Lady of the Assumption Church, with help from the Fathers of the Sacred Hearts.

Figure 2.5 An altar near Cidade Velha (old city) lies in front of the Atlantic Ocean.

In the late 1800s, Cape Verdean Protestant evangelists began to convert other Cape Verdeans in the United States. Manuel Ricardo Martin founded the first Cape Verdean Protestant church in the United States in Massachusetts. After living ten years in America, João José Dias returned to the islands in 1901, with the intention of establishing Protestantism. He tried diligently for 35 years but was criticized, jailed and beaten. He was accused of leading people astray from the Catholic religion, and thus leading people away from the Portuguese government and its ideologies. Nevertheless, he succeeded in planting the seeds of Protestantism. In 1936, Reverend Everette D. Howard from the Church of the Nazarene went to Cape Verde to continue the work begun by João José Dias. More people converted to the Protestant faith. Today, 10 percent of the population is Protestant, half of which are members of the Nazarene Church. Other Protestant groups are Baptists, Seventh Day Adventists, and Mormons from the Church of Jesus Christ of Latter Day Saints (Almeida, 1997b).

Jewish Cape Verdeans

Although there are no Jewish synagogues, rabbis or practicing community left today in Cape Verde, Jewish religious and cultural identity and history was important to Cape Verdean history and remnants of their significance is apparent throughout the islands. In 1492, the Spanish Inquisition spread its ideas of prejudice and anti-Semitism, torturing, killing, and evicting Jews from cities and countries in Europe. These practices reached Portugal where Kings João II and especially Manuel I exiled thousands of Jews to São Tomé, Príncipe, and Cape Verde (Lobban, 1996; Werlin, 1995; Almeida, 1997b).

Jews were among the first settlers to arrive on the previously inhabited islands. They first landed on the island of Santiago. Jewish immigrants were rejected by the larger Cape Verdean community and were restricted to living in a small community in Praia, one of the more populated towns of the time. In spite of their exile or convict status, they were permitted to engage in a small amount of trade in specific businesses such as hides and coffee. After 1548, Jews settled in the Northern island of Santo Antão where there is still evidence of their community in the name of the town Sinagoga, and in the Jewish cemetery in the village Ponta da Sol (Lobban, 1996; Werlin, 1995).

In the 1850s, Moroccan Jews arrived in Cape Verde, especially on the islands of Boavista and Maio in order to trade hides. Several Jewish cemeteries have survived with Hebrew inscriptions on gravestones, mostly of Moroccan born Jews and many Jewish family names exist today throughout the islands (Almeida, 1997b; Lobban, 1996; Werlin, 1995).

Independence from Portugal

Ideas of independence began to expand in the minds of Cape Verdeans as they did all over the world in the middle of the twentieth century (Luz, 2007). Cape Verdeans came to play a leadership role in the anti-colonial movements providing the spark for independence in other Portuguese African colonies. The efforts of political activists in Cape Verde and other nations held by the Portuguese undermined Portugal and led to both the overthrow of Portugal's fascist dictatorship and the establishment of former colonies as independent states (Fajana & Anjorin, 1979). Cape Verde gained its independence in 1975 with Amílcar Cabral, a Cape Verdean raised partially in Guinea-Bissau, as one of its primary leaders.

In 1951, Portugal attempted to head off the independence movement in Cape Verde by changing its status from a colony to an overseas province. But their policy change did not stop the Cape Verdeans seeking independence. Cabral had been a part of a student organization in Lisbon which established an African Studies program there. During their university studies, Cabral and other students wrote about ways African people could re-Africanize themselves and how to denounce the colonial powers that had stolen their African identity. Although their activities had enormous practical geopolitical consequences, the focus of their work, especially initially, was highly academic establishing university programs and debating intellectual questions about culture and ideology. Their protest began through critiques of literature which they believed was dominated and distorted by colonialism (Fage & Olivier, 1978).

In 1956, Cabral and a group of Cape Verdeans and Guinea-Bissauns organized the clandestine African Party for the Independence of Guinea-Bissau and Cape Verde (PAIGC). They demanded economic, social, and political reform and founded the independence movement in both Guinea-Bissau and Cape Verde. When their demands were not met, the PAIGC obtained support from the Soviets and began an armed rebellion against Portugal in 1961. In 1974, the PAIGC and Portugal signed an agreement providing for a transitional government composed of Portuguese and Cape Verdeans and the first elections were held in 1975 (U.S. Department of State, 2008).

Cabral was executed in the early days of the revolution in Guinea-Bissau in 1973, but his ideas continued to shape the style and content of politics and economics in Cape Verde (Robinson, 1981). Cabral (1970) argued that culture was a critical focal point for political struggle. He maintained that colonialists and imperialists who sought to subjugate a people strove to suppress their culture as a way to keep them from developing and challenging the rule of the dominant forces. But, he argued, culture was also a focal point for revolution and a critical site around which revolutionaries must organize to resist and confront those forces. Cabral wrote,

> Just as happens with the flower in a plant, in culture there lies the capacity (or the responsibility) for forming and fertilizing the seedling which will assure the continuity of history, at the same time assuring the prospects for evolution and progress of the society in question. Thus it is understood that imperialist domination by denying the historical development of the dominated people, necessarily also denies their cultural development. It is also understood why imperialist domination, like all other foreign domination for its own security, requires cultural oppression and the attempt at direct or indirect liquidation of the essential elements of the culture of the dominated people.

From 1975 until the 1990s, Cape Verde was ruled by the revolutionary party, PAICV which had toppled the colonial rule. Responding to pressure for a participatory, multi-party system, PAICV ended one-party rule in 1990. The first multi-party elections were held in 1991 with the opposition party MPD taking control, and again winning the election in 1996. Power returned to PAICV in 2001 and continued again in 2006 (U.S. Department of State, 2008; Chabal, 1996). United Nations documents declare that in 1991, principles of democracy, free market reforms, and socioeconomic progress have commenced in Cape Verde (United Nations, 2006b).

Food and Famine

Despite Cape Verde's tragic history of slavery, famine, and colonialism, today it looks ahead for ways to build the economy through better land management, and developing the fishing industry and shipping services and especially tourism. One aspect Cape Verde officials and business people hope will draw tourists is its food. The islands' easy access to the sea, the cultures of the slaves imported from West Africa, and the nation's status as a former Portuguese colony all contributed to its cooking customs. Slaves from West Africa cultivated bananas, mangos, and papayas, while the Portuguese transported livestock into the country. Together they grew corn, pumpkins, and cassava brought from the Western Hemisphere and Africa on Portuguese trading ships. Today's national dish, *cachupa*, is a mix of corn, beans, and fish or meat. For Cape Verdeans around the world, *cachupa* symbolizes the country (Cape Verdean Foods, n.d.).

But yesterday's difficulties persist and problems in agriculture may hinder the tourist industry. Fresh water remains a scarce resource. Farmers are challenged in their attempts to get water into the soil. Reforestation, catchment dams, and drip irrigation are projects that are proving somewhat successful. Finding water to drink is a problem as well. Some islands have desalination plants. Others drill deep wells or use windmills to pull water out of the ground. Arable soil is another problem. Wind and water erode the soil and farmers terrace up the mountains at increasingly great heights. Finally, insects such as grasshoppers and millipedes devour potatoes and carrots while huge numbers of goats and especially locusts demolished the already scarce vegetation in 1994 and again in 2004 (Irwin & Wilson, 1998; IRIN News, 2004).

Droughts and famines continue to curse the country. A 10-year drought stopped briefly in 1978, but returned in 1979, 1981, and 1983 when the drought was so severe that almost all crops were lost. In 1984 heavy rainfall caused devastating floods, which were followed by drought in the 1980s and 1990s (Africa—South of the Sahara, 1997). In June 2002, with famine threatening 30,000 citizens, the Government made its first appeal for emergency food aid in more than 20 years. The UN World Food Program responded with a request to international donors for $1.2 million in emergency aid (*Europa World Year Book*, 2003).

Famines are not caused only by drought. Food shortages also are a byproduct of agricultural practices that have historically and continue to be organized around export products and crops that are inappropriate to the environment. In addition, even agriculture that is producing for domestic needs is focused on nonfood products. For example, today most of the farms that use irrigation water raise sugarcane that is often used to produce alcohol spirits, *grogue* (Langworthy & Finan, 1997). Furthermore, agriculture in Cape Verde was historically organized by the Portuguese crown that established maize as the staple crop. Maize, however, is highly inappropriate to the soil and rainfall of the natural environment (Moran, 1982). In addition, many farmers

Figure 2.6 Only a small part of the population has running water. Everyone else collects water at these community water channels every morning and evening. This picture was taken on São Nicolau Island.

in Cape Verde work within a sharecropping system which requires that they bear all of the costs of production and give 50 percent of their harvest to the landowner (Bigman, 1994; Langworthy & Finan, 1997).

Famine is indeed related to the poor soil and low precipitation of frequent drought years. Famine is also a result of the social organization of agriculture, especially the production of crops that are geared toward the needs of people outside Cape Verde who consume the goods and the wealthy people inside the nation who produce the products for sale. The needs of the people of the islands for food are ignored and hunger results.

The Sea

Cape Verdean food, music, and poetry are all connected to the ocean. The ocean symbolizes isolation from the rest of the world, as well as a deep, sad longing—*sodade*—for family and friends who have migrated to new places (Almeida, n.d.). Not surprisingly, fishing on this island nation has historically been a major activity, but the fishing industry is currently struggling.

Today, almost 3,000 Cape Verdeans work in artisanal fishing and the same amount in industrial fishing. There are about 100 men who are lobster divers—either as free divers or on commercial lobster boats. Artisanal fishermen are called such because "their success depends on their mastery of the art and craft of fishing" (Almeida, 1996).

From 1985 to 1991, artisanal fishermen brought in somewhere between 4,000 and 7,000 metric tons of fish; they catch only for the domestic market. In addition, they produce more than 65 percent of the national production of fish (Almeida, 1996).

There are approximately 20 fishing villages and over 60 docks along the coasts. Each fishing village has skilled net makers, motor repairmen, and carpenters. Learning to fish is seen as a rite of passage for young boys. They are taught by their fathers or other skilled men. Women cut and clean the fish, salt, or dry them (Almeida, 1996). Fishing, like the whaling industry that was central to Cape Verdean life centuries ago, is a risky business. Lobster divers especially face the threat of decompression accidents, shark attacks, and running into giant manta rays. The waters around the islands are typically jarring with strong winds. Fishermen are commonly caught in the rough waters, wrecking their boats on rocky shores. Ship wreckage from the past 500 years are scattered on the ocean floor throughout the islands (Almeida, 1996).

If a fisherman dies while at sea, villagers organize a day of mourning: no one fishes on that day, men play *mornas* [a slow piece of music] on guitars and violins, a priest leads the ceremony, and *cachupa* is prepared as well as *grog* [national alcoholic drink] (Almeida, 1997a).

Despite its risks, fishermen also have a special relationship with the sea. The daily routines and rhythms of the work provide solace in a place that regularly faces famine, drought, and massive migration. Raymond Almeida (1996) interviewed local fishermen and asked how they distinguish between a good or bad fishing day. Fishermen reported that they read the countless signs of nature throughout the days and seasons. He summarized their connection to the sea:

> What phase is the moon?
>
> Which is the best tide to sail on to be at the fishing grounds when the tide is flowing?
>
> Do clouds hide rain to keep the ocean surface and my face cool?
>
> Is there enough of a breeze to make sail?

> Does the wind come from the northeast so the day will be dry or from the south whence most storms blow?
>
> Do whitecaps crown the waves or will the sea be a cradle with a gentle rolling motion?
>
> What fish at this time of year and what bait? Is that rippling over there a school of bait fish?
>
> If I am going out for surface-feeding fish, will they take to bait in the morning and late afternoon rather than in the heat of midday? If I am going for bottom fishing, the time of day won't matter as much as the tide and currents.
>
> Could that passing pod of dolphins signal that tuna swim deep beneath them?
>
> Do those sea birds diving tell of a school of tiny bait fish?
>
> Or maybe a breaching grey whale?
>
> Is the water so warm that I am more likely to encounter sharks today?
>
> And finally, will I have to race to port in order to arrive early enough to sell my catch in the local market place?
>
> Or will there be demand for my fish at the hotels and restaurants regardless of what time I get back?
>
> As a last resort will I be able to sell my catch at the tuna factory if nobody else will buy it? (Almeida, 1996)

Cape Verde's productive fishing grounds have not been overlooked by outsiders. Foreign investors, exporters, and middlemen negotiators, however, have caused problems for local fishermen. They tend to partner with the government, overlooking the needs of artisanal fishermen and their villages. Investors raise local prices with the intention of seizing as much catch or profit as they can. The price of lobster, for example, is not affordable to many Cape Verdeans, even for the divers who caught it themselves (Almeida, 1996).

The importance of artisanal fishermen to nourishing the country has been ignored or underestimated by government officials. Overfishing and overdevelopment with the objective of exporting hurts the produce of the artisanal fishermen that provides more than 65 percent of the national production of fish as well as 75 percent of protein eaten by the citizens of Cape Verde every year. Meetings of government officials and fishing interests include commercial boat owners, cannery workers, lobster exporters, and international companies to discuss issues of boat breakdowns and repair, bank and loan issues, international market demands and fish development. Artisanal fishermen, however, are usually not represented (Almeida, 1996). The problem of excluding the fisherman who provide for local food may be exacerbated as globalization is intensified in Cape Verde.

Because the country recently joined the WTO (July 2008) and has borrowed millions from the World Bank, many Cape Verdeans fear that the objective of economic reform will take precedence over cultural sensitivity and protection, especially of the artisanal fishing communities. They have witnessed the experience of artisanal fishing in the Canary Islands where the World Bank, other international lending agencies, and aid donors have neglected to look after local culture. Huge economic, fishing, tourist, and business developments and reforms have transpired there but at the cost of disregarding the local culture, customs, and ways of life. Cape Verdeans hope new developments in these areas will not disrupt their artisanal fishing community as they have in the Canary Islands. They argue that both commercial fishermen and artisanal fishermen contribute an important part of the economy and culture and should not be overlooked (Almeida, 1996).

Cape Verde has a long history of exploitation and difficulty resulting from its connection to other more powerful nations first through the slave trade, then through colonialism and today through the forces of economic globalization. Its economic life, spiritual life, and even its food are shaped by outside forces as Cape Verdeans have attempted, and continue to attempt to survive and flourish. Our research seeks to explore the contemporary scene within this context, especially the experience of women, as it is viewed through the eyes of Cape Verdeans.

Reflections on Chapter 2 Research Methods

This study is a critical ethnography which is a type of qualitative research method. In the next chapter we will describe in detail the methods we used to conduct our study, but here we would like to provide you with a list of the characteristics and aims of qualitative research so that you can see how this chapter fits into that methodology. Immy Holloway (1997) offers the following items to describe what qualitative research must do:

- Focus on everyday life of people in their natural setting.
- Maintain primacy of data.
- Because the research is context-bound, researchers need to immerse themselves in that context and explicate the setting and surroundings.
- Make an emic perspective central, that is, priority is given to the view of the people involved—their perceptions, meanings, and interpretation.

- Provide detailed descriptions.
- Acknowledge the equal relationship between researcher and researched.
- Proceed through data collection and analysis simultaneously (p. 5)

Chapter 2 begins to address some of these requirements. It provides a detailed description of everyday life in Cape Verde and presents a picture of the context in which the people live. This kind of detail can only be accomplished by long term immersion in that context. Ethnographers must spend an extensive amount of time with the people whose lives are being studied. Carter, for example, lived and worked in Cape Verde for three years. Before beginning the study and throughout the research, she learned the language, participated in weddings, funerals, religious festivals, and dance parties. She became friends with the neighbors in her community and her colleagues and students at the university where she was teaching. She often observed and frequently participated in daily rituals of fetching water, pounding corn, waiting for the bread woman to pass by and without fail, waiting for the local bus to come.

Ethnographers collect data through participant observation, conversation, and interviewing. Ethnography is full of rich, in-depth detail. It is a method historically associated with anthropology, but it is also used in other disciplines including sociology, political science, history, education, business, linguistics, folklore, and health. This methodology is grounded in the experiential learning approach—which asserts that most of us learn by doing. When we study another culture, understanding and interpretation depends on oral communication and shared participation in cultural experiences (PERCS, n.d.). This chapter is based on the shared experience of the researcher and the researched.

We have also included ten photographs in this book, some of which are shown in this chapter. All the photos were taken by Carter. Photographs are used in ethnographic research in order to present aspects of community and daily life which might not be evident to the reader from the written document alone. Photographs further show people within their natural settings. We can see what "people do and the physical contexts in which it is done" (Schwartz, 1989).

In presenting photographs for this book, we have attempted to create "a record about culture" (Worth, 1980). This record about culture attempts to accurately document a substantial amount of concrete information and allows for thorough examination by readers and viewers (Prosser, 1992). In this book, our record about culture provides information on what men, women, and children do in their everyday lives, including the clothes they wear, the dances they perform, the physical structures of churches, and the physical environment of the place where they live such as the sea, volcanoes, and the banana fields. This allows us to further disclose and explicate the meaning of everyday life for Cape Verdeans.

In some research settings, the camera may cause anxiety in a community. Taking photographs of those not interested in being photographed could be invasive, possibly damaging the rapport of the researcher with the group being studied (Prosser, 1992). Carefully done, however, researchers report that using photography often helped them gain acceptance and entrée in the community and served as a helpful way to gather important data (Prosser, 1992). Collier and Collier (1986) recommend taking photographs on the first day of meeting the community, serving as a "social can opener."

One anthropologist who used photography explained,

> The camera itself became an important means of entering into the social life of the community, allowing me to engage in understandable, task-oriented activity in the course of observation. My picture-taking provided residents with an obvious reason to start up a conversation, and the longer I made photographs, the more people I met. I was able to move from photographing the environment to photographing public events as my contacts with community members multiplied. (Schwartz, 1989)

Carter's experience was a bit different. She didn't need to use the camera as a way to enter the community; she was already a member of the community when she began taking snap shots. She knew that the photographs might be used in her research, so she explained to each person and group of people that the pictures she took were a part of the research she was conducting, and their picture might end up in a book one day. Most everyone was compliant with having their picture taken, while a few politely refused.

Throughout this book, we present the results of our ethnographic and qualitative interview data, including the use of photographs. This chapter also draws from other text sources to expand the picture of Cape Verde and its history. Chapter 2 sets the stage for helping the reader understand the interview data by explicating the setting and surroundings. A description of slavery and colonialism was given in this chapter, for example, in order to understand participant's description of solidarity and resistance in the chapter on *batuku* dance. We also mention Amílcar Cabral and his activities in Cape Verde as his ideas are important to the history of Cape Verde, especially the struggle against colonialism which is mentioned in the interviews. In addition, Cabral's ideas are critical to our understanding of the importance of culture in the resistance movement in Cape Verde and are further developed in the last chapter when we ask "what does this all mean?"

In the reflection on methods section in chapter 1, we mentioned that one critical piece of a research project is a literature review. Chapter 2 serves as part of the literature review for this study because it addresses the question of what have other scholars written about our subject. This chapter tells us what others have written about Cape Verde in general. It provides a review of the literature on Cape Verde's history and current economy. Now that we know a little about what other scholars have written about Cape Verde, we will move on in the next chapters to explore our research question of how this context affects women's lives and how they are responding to the problems they face.

This chapter is only the first of the places where we incorporate a review of the literature. In each of the chapters reporting data, we also ask what other scholars have found. And in the final chapter where we try to come to some theoretical conclusions based on our empirical findings, we review some of the literature written by theorists on issues regarding the connections among gender, culture, and resistance.

3

Insiders and Outsiders Exploring Cape Verde

Our study of Cape Verde was done using two forms of qualitative methods, a thematic analysis of critical ethnography and critical discourse analysis. In the previous chapter we outlined the characteristics of ethnography. Here we will elaborate on that description. In addition, we will spell out the differences between ethnography and critical ethnography and the ways in which our research fits into the latter type. We will also provide a review of critical discourse analysis and the way we used it as a tool for understanding the data on Creole.

Methods of Data Collection

Katherine Carter came to Cape Verde as an English teacher in 2001. A lifelong journal writer she made notes on what she was seeing and hearing and learning but as she got to know the people in the community she decided to begin to systematically collect data in 2003 in order to record her observations and document the voices of the people who were becoming her friends. She invited Judy Aulette to work on the project with her to help conceptualize and construct the research project and to analyze the data and write about it.

Observations were made in Praia, the capital city, over a period of twelve months. A qualitative approach was taken to this study in the belief that quantitative measures would not provide valid indicators for understanding language, culture, and women's experiences and relationships, especially since the researchers and the researched came from such different social groups (Stanley & Wise, 1983).

Qualitative research began to develop in the nineteenth century as the approach used by British anthropologists in their studies of people from cultures in other regions of the world. In the early twentieth century, qualitative methods were further expanded and came to be associated with the Chicago School in the United States because it was practiced by scholars and activists from the University of Chicago in their studies of neighborhoods in Chicago. In 1978, the first scholarly journal dedicated to qualitative methods, *Qualitative Sociology*, began publishing. Since then, qualitative methods have gained increasing acceptance in many fields, including those that had formerly paid attention only to quantitative research. For example, in 1990, the World Health Organization published an overview of concepts and methods of qualitative research (Holloway, 1997).

Globally, qualitative methods have surpassed quantitative methods in its use by social scientists. The majority of British scholars, for example use qualitative methods in their research and only one in twenty articles published today in Britain report work that uses quantitative analysis (Payne et al., 2004). This proportion is similar in other areas of Europe (Alasuutara et al., 2009).

Put most simply, qualitative researchers collect and analyze words and images in contrast to quantitative researchers who gather information that can be quantified and who use numbers to analyze issues. Qualitative researchers also have a preference for naturally occurring data (as opposed to data generated by manipulated situations such as experiments). Qualitative research, furthermore, makes meaning central, especially meaning from the point of view of the participants (Flick, 2006). Holloway (1997) writes, "Qualitative research is a form of social inquiry that focuses on the way people interpret and make sense of their experiences and the world in which they live. Qualitative research looks at the everyday life, social interactions, and language of people" (p. 1). As these are the topics about which we are most concerned, this approach is ideal for our interests.

Ethnography is a particular form of qualitative research that involves the researcher in the field, the social context of the study, for extended periods of time. The ethnographer collects information through interviews and participant and nonparticipant observation. The central goal of the ethnographer is to find the emic meaning of events, ideas, and relationships. This means that the ethnographer strives to describe and interpret the lives and culture of the people being studied from their point of view, within their culture (Bransford, 2006).

This study is a particular kind of ethnography that is called critical ethnography (Thomas, 1993; Madison, 2005). The word critical here means critical of the injustice and inequity of the status quo (Johnstone, 2002). This idea comes from Karl Marx's (Marx & Engels, 1998 [1845], p. 570) statement which so clearly summarizes his view, it was inscribed on his gravestone, "The philosophers have hitherto only interpreted the world, in various ways; the point is to change it."

Critical ethnography goes a step further than what might be called mainstream ethnography. "What is important for the critical ethnographer is that the probing of the subjects' meaning is not the end of the story…it [the subjects' meaning] is not independent of structural factors" (Harvey, 1990, p. 120). The critical ethnographer seeks to find not only the meaning from the participants' point of view but how political structures shape that meaning and how participants help maintain those structures or challenge them. In other words, the critical ethnographer seeks to understand the issues from the point of view of the participant, and, within the political context of the participant.

Lee Harvey (1990) writes, "At the heart of critical social research is the idea that knowledge is structured by existing sets of social relations. The aim of a critical methodology is to provide knowledge which engages the prevailing social structures. These structures are seen as oppressive structures.…Critical ethnography attempts to link detailed analysis of ethnography to wider social structures and systems of power relationships in order to get beneath the surface of oppressive structural relations" (p. 2). Critical ethnographers, therefore, must not only elicit and explain the participants' experience, they must also disclose, discuss, and critique the oppressive political structures that shape life in the community under study. They begin with the reflections and ideas expressed by the participants and then move to finding where those images and words fit within the political relationships that are intrinsic to the lives of the participants. In the case of Cape Verde, these political arrangements include a legacy of slavery, colonialism, and famine; a contemporary economic and political position in the South in a globalized world dominated by the North; and a society that is stratified by gender in which women are often the exploited subjects of men in their personal and public lives.

Critical ethnographers not only differ from mainstream ethnographers in approach. They also differ in regard to intentions. Critical ethnography "seeks to do more than just describe and interpret culture and cultural phenomenon. Critical ethnography seeks to change it. Specifically critical ethnography seeks to make visible those covert structures of power and oppression that subtly and yet forcefully 'construct and limit choices, confer legitimacy, and guide our daily routine' (Thomas, 1993, p. 3)" (quoted in Bransford, 2006, p. 178).

In this study we seek to listen to the voices of the people of Cape Verde as they talk about their everyday lives. Furthermore, we seek to find the meaning of their ideas within the power structures that shadow their lives, the global economy, the Cape Verdean government, and the gender systems that shape women's personal relationships with men. And, our purpose in conducting the research is to determine ways to alter the status quo and promote equity and justice in our globalized world.

Gathering Data

Three methods of data collection were used for this research. First, 40 students who agreed to participate filled out a questionnaire on topics such as dance, language, and economics, which asked them open-ended questions (see appendices A, B, C). These students provided written answers to the questionnaire that we later discussed with them.

A second set of data was collected by teams of multilingual Cape Verdean students (Creole, Portuguese, and English) who worked as research assistants. They interviewed 36 local women about their everyday lives and their relationships with work, families, and men. These questions are listed in appendix D.

Third, we distributed 50 questionnaires to women attending a women's rights conference in a rural area in order to understand the problem of violence. These answers were written in response to the open-ended questions listed in appendix E.

Ethical Considerations

All research with human subjects must be aware of important ethical considerations and must closely follow certain protocols to protect research participants. Following the Second World War and the discovery of the horrific "experiments" conducted by the Nazi's, the Nuremberg officials developed a set of ethical principles by which all researchers must abide. In the middle of the twentieth century, the disclosure of unethical and inhumane study by the U.S. Public Health Department in the "Tuskegee Study of Untreated Syphilis in the Black Man" that took place between 1932 and 1972 further solidified the need to formalize legal restrictions on researchers (Center for Disease Control, 2008). The World Health Organization (WHO) established in 1964, and later extended in 1975 the Declaration of Helsinki that outlined ethical principles for medical research involving human subjects which has been extended to all scientists conducting research on human subjects. Every nation has created its own ethical standards in reference to these international declarations. In 1974, the National Research Act was signed into law in the United States, creating the National Commission for the Protection of Human Subjects of Biomedical and Behavioral Research. The commission was charged with identifying the basic ethical principles that should underlie the conduct of biomedical and behavioral research involving human subjects and developing guidelines which should be followed to assure that research is conducted in accordance with those principles (National Institute of Health, n.d.).

Social science research is usually not as risky or potentially damaging as conducting medical experiments on patients. We do face risks, however, and we took care to protect our participants. In gathering our data we followed ethical guidelines providing participants with the following rights: no harm would come to them as a result of participation; they were informed of the purpose of the study; they consented and participated only on a voluntary basis; and their participation and responses were kept anonymous and confidential (Holloway, 1997).

Breeches of privacy and confidentially are probably the most potentially harmful aspects of this research. We asked questions of students and women and received responses surrounding sensitive issues: national and local politics, inequality, infidelity, community networks, and work situations. If the responses were unintentionally traced back to the participant, relationships might be damaged and broken, and trust betrayed. Therefore, we took important steps to protect the privacy and identities of the participants. We have used pseudonyms in this book to identify both the women's and the student's comments. And we ensured that personal information (e.g., names of family, places of work, names of schools) could in no way be traced back to them. Our research was also approved by the human subjects review board at Aulette's university, which deemed our work ethical and safe, foreseeing no harm would come to our participants.

Problems with Collecting of Data Cross-Culturally: Telling (What Is) the Truth

The task of collecting valid information in the field is a central challenge for all researchers. When research is conducted in another culture, the problems are even more pronounced in regard to differences of history, social context, nation, class, language, ethnicity, values, and practices. One challenge is communicating across cultures. When deciding to conduct research, all researchers face the problem of examining lives of people who are different from themselves. These challenges include making contact with the people being studied, finding the right people with whom to speak, asking the right questions, and interpreting the answers and views of the participants. In addition ethnographers must deal with the problem of distorted information, or in some cases, informants who choose not to tell the truth (Baca Zinn, 2001).

At least two factors seem to play a role in why informants purposely do not provide an ethnographer with correct information. Informants may choose to at least obscure the truth if some other value is being challenged. Informants may also choose to avoid the truth because they believe that exposing information about oneself or others is not acceptable behavior.

Ashkenazi (1997) found examples of both of these in fieldwork he conducted among Ethiopian immigrants in Israel who sometimes gave him false information. His informants were most likely to avoid the truth when they felt defenseless or exposed, and they believed that lying was a way of protecting themselves. One example he wrote about regarded the issue of marriage and divorce. The seemingly simple question of "are you married?" was fraught with surprising complexity and likely to elicit incorrect information.

In Israel, early marriage is not possible, and is moreover, viewed with repugnance by much of the population. Divorce in Israel is also controlled by a set of rabbinical rules, which are often difficult and burdensome. Many Ethiopians may have married very young and/or they may have not obtained a legal divorce although they no longer felt themselves married. Failure to abide by the social and legal rules in Israel can have severe consequences not only for oneself but also for the children of individuals who had remarried after an "improper" divorce. Ethiopian immigrants, aware of this problem, avoided discussing or if necessary, lied, about marital status, previous marriages, or divorces (Ashkenazi, 1997).

In this case, the informants chose what they perceived to be the "lesser of two evils." Some cultures might prioritize "telling the truth." In fact, the ethnographer reveals the centrality he/she gives to "seeking the truth" and "reporting the truth" in his/her research. The Ethiopians in this case, however, show their concern for giving priority to other values over "telling the truth." They are more concerned with protecting their reputation and avoiding stigmatization of themselves and their children. This kind of self-protective behavior is especially evident among people in minority communities who encounter questioners from dominant communities (Baca Zinn, 2001).

In some cultures "telling the truth" may be seen not only as a less important value than others; "telling the truth" may be seen as unacceptable if it exposes information that is considered improper (Condon, 1986). In another example, Ashkenazi (1997) found in his research in Japan, "my Japanese informants would probably have been highly embarrassed had I been 'honest' about everything I did and thought. Keeping one's thoughts to oneself and preserving, by almost any means, a placid set of social relations is valued much more highly."

Both of these examples show the ways in which informants can choose to shape their response to an ethnographer in a way that is consistent with their beliefs about what should be shared, perhaps especially with outsiders. Informants can also, however, convey information that is not fully understandable or easily misunderstood not because of their own decisions about what is proper or improper, but because of language differences between researchers and researched.

Lost in Translation

Conducting research across language lines further exacerbates the problems of researching across cultures. Differing languages between researchers and researched present an important barrier to understanding. The difficulties include both literally understanding the words as well as understanding the meaning of the words within the context of the culture. Cross-cultural researchers have developed strategies for crossing language lines (Dunnett et al., 1986).

First, they argue languages cannot be translated word-for-word. All languages have idioms, proverbs, and expressions that hold implications that differentiate beyond the meanings of the individual words themselves. Second, the tone of a person's voice (the intonation pattern) holds meaning. Each language has different "harmonies" or patterns of intonation. Equally, expressed language varies in the extent to which loudness/softness is used by individuals. Third, body language is important as gestures and movements in a culture hold meaning. Finally, all cultures have taboo topics. These taboos are reflected in words and topics that one should and should not say in specific circumstances.

As a way of confronting all of these problems associated with cultural and linguistic diversity between researchers and researched, we decided one solution was to involve Cape Verdean students in the research: to help overcome some of the outsider barriers; to address the problem of obtaining valid information; and to help conceptualize the issues from the point of view of insiders (Humphrey, 2007; Braithwaite et al., 2007; Bozalek, 2004).

Carter was teaching students who were enrolled in a university program. We recruited research assistants to help with the study from among these university students. We asked them to participate in gathering and analyzing data because they were members of the community and would be able to hear the participants better than we would (Humphrey, 2007; Braithwaite et al., 2007; Bozalek, 2004). In addition, we invited them to work with us because of their language skills, as they were fluent in Portuguese, Creole, and English. Furthermore, we chose them because of their knowledge about their communities and their leadership role in Cape Verde as a highly educated stratum.

These university students would soon enter a professional stratum that was knowledgeable about their country, concerned about its future and called upon to play leadership roles in the community. They comprise an important segment of Cape Verdean citizens because they are among a relatively small proportion of people who continued their education past secondary school. They, therefore, represent the most academically successful citizens in Cape Verde and play a central role in shaping public understanding and public opinion.

Students conducted interviews with 36 women. The interviews were conducted in Creole and were later translated into English. Students submitted their first interview for feedback, before conducting interviews two and three. After students completed the interviews, they were transcribed and analyzed using a qualitative analysis software program, Ethnograph 5.0. In addition to the transcriptions of the interviews, students made impressionistic notes and recorded their perceptions of participants' responses and feelings. Students developed themes from these primary data, conducted library research to find pertinent literature, and put it all together into a final report which we then used as data in addition to the transcripts of the interviews themselves.

Difficulties Students Faced

We believe that we collected valuable data that reflects an accurate snapshot of life for many Cape Verdean women. Some difficulties in data collection, however, are worth mentioning. Students encountered problems that reflected both their outsider relationship to the participants and their outsider relationship to us. Students experienced three common dilemmas: inadequate questions; cultural sensitivities and values; and translation difficulties. The first problem was the questions and wording of the questions themselves. Before beginning the interviews, Carter reviewed the questions with the students and made changes based on their suggestions. During the interviews, however, the students discovered that additional questions or differently worded questions were needed to get a complete picture from the participants.

Adam, for example, asserted that he believed that some of the questions missed the mark but he was initially hesitant to tamper with the list. His comment reveals the kinds of relationships of power that can get in the way when outside investigators try to work with inside research assistants. He assumed that we knew what we were doing and that his task was just to follow through. When he had to write his own report based on the data, however, he mentioned that the questions were not as good as they should have been to obtain the most valid data. He explained.

> [For] the first interview I conducted, I limited myself to asking only the questions from the list. This was because I was thinking we would reach the desired outcomes of the work only with the questioned we had asked. However in the results I came to see that some additional questions were needed to get the clear message from the participants. (Adam)

The second issue students faced surrounded the issue of cultural sensitivities, values, and obtaining candid answers. To address this difficulty, the students created strategies to probe in ways that provided more valid data. These included using more colloquial language, repeating questions, and asking in more indirect ways. But students also explained that in some cases they chose not to pursue fuller answers because they did not want to push when women were hesitant to elaborate. For example, students said some of the women interviewed did not want to talk in detail about their relationships with men. And, consistent with ethical treatment of human subjects, we assured the women that they did not have to answer any questions they preferred not to.

The issue of truthfulness emerged as particularly significant in asking women participants to talk about their relationships and partner betrayal, which they considered sensitive. Some students experienced no problems during the interviews with these sensitive topics although they acknowledged that they might have anticipated problems because they are issues about which women might feel uncomfortable. One student said:

> My participants showed an amazing easiness during the interviews and one even suggested another friend whom I should talk to. (Zach)

However, most students encountered difficulties. Jorge, for example, explained that he had difficulties as an insider because of his knowledge of the sensitive nature of the questions. He said:

> I felt nervous and uncomfortable at the beginning of the interview....I found myself in a dilemma because I didn't know how to start the interview. I didn't know what the women's reaction would be towards my questions...I wanted to be very careful in order to avoid misunderstandings and hurt feelings. (Jorge)

Other students mentioned their own feelings but their emphasis was not on their own discomfort but that of those they interviewed. Students, such as Adam, felt that talking about sensitive topics and asking questions that might be misunderstood by the interviewees made his participants feel uncomfortable. He also shows how respect for his interviewees required that he not press too much if they seemed hesitant to answer the questions. He explained,

> Some of the questions were unclear to participants, or had to be repeated many times. I found that the nature of the work I conducted did not allow for further pushing or probing. I found that this work was sensitive, dealing with people's feelings, beliefs, and attitudes. For these reasons during the interview I had to sometimes respect my participants without probing. (Adam)

Charlie shared Adam's observation of discomfort among his interviewees. Instead of backing off from the questions he tried to negotiate between the outsiders (Carter and Aulette) who had asked him to conduct interviews and the members of his community he is interviewing. He created a strategy for overcoming the sensitive barriers:

> All my participants answered my questions. However there were some questions which I had to create a favorable environment to ask, or in other words to ask the question in a less-direct fashion. Some questions that were sensitive related to women's partners and the concept of betrayal. Sometimes I needed to clarify my questions because one of my interviewees didn't understand the term "family planning." Other times I needed to ask the same question in a different form, because interviewees did not want to respond at first. For example, when I asked Helena why her husband and she separated, she became shy and gave a little smile. I reworded my question and reminded her that her name wouldn't be included in the paper. (Charlie)

Charlie changed the language in order to conduct an effective interview. He chose to be less direct as a way to create a more comfortable interaction and persuade the interviewee to answer his question. Ellen also mentioned that the questions were too direct, but she did not try to make them less direct. Instead, Ellen used repetition as a strategy.

> Some of my participants didn't give me a direct answer about the questions related to their relationships with men. I suppose they didn't feel comfortable to talk freely about their relationships. Some questions I asked were not clear enough, and sometimes I had to repeat the questions. (Ellen)

Joe, like Charlie, changed the language that was used in order to get more information. He did not become more subtle as Charlie had. Nor did he repeat himself as Ellen had. Instead, Joe chose to use local language:

> My interviewees had trouble understanding the questions "does your partner live up to your expectations?" and "does he provide for you emotionally?" I had to ask them in more local terms, and the last question was omitted because I couldn't explain to them exactly what "emotional support" means. (Joe)

In addition to the hesitancy to answer questions and the apparent inability to understand the questions being asked, some interviewees may have chosen to not tell the complete truth and student interviewers warned us that the data might not be valid. Drew revealed his suspicions about the sincerity of the participants:

> One issue I thought about related to the sincerity of the interviews. Were they telling me the truth or hiding something? (Drew)

Adam was less suspicious of the motives of people who did not answer questions entirely honestly. He explained his belief that their behavior was justified and rational:

> I could see that in some issues such as relationships, people would give me the ideal information just to save face. This is because throughout my interview I could feel participants contradicting themselves when I later asked them similar or related questions. (Adam)

Besides feeling uncomfortable, some interviewees may not have understood the questions because they were written from the point of view of someone outside of their culture. Joe and Charlie mentioned the terms "family planning" and "emotional support" were confusing to their participants. Cape Verdean men and women plan for families by using birth control pills and condoms. And they care for the emotional feelings of their partners. But the terms "family planning" and "emotional support" are not the way these activities are described. We mistakenly assumed these terms are common and known in all cultures. Both students and participants made us aware of these important gaps in the tools we had created to conduct this research.

The final problem students experienced was bridging the language gap in the research. This included translating the questions from English into Creole, and the responses from Creole into English. It also demanded translations of the meanings of special words and phrases, for example proverbs which serve as a short way of conveying information in Cape Verdean discourse. Although the students were fluent in Creole as well as English and were familiar with the proverbs and Carter became fluent in Creole, finding ways to capture the exact meaning and translating it into English was difficult and not always possible. Charlie, for example, described his difficulties in translating:

> I wrote word by word what the participants said. I wrote what they said in Creole and then translated it into English. However some changes

> occurred here, because in some aspects it is not possible to translate literally what the participants had said. (Charlie)

Jorge specified what part of the interviews—the proverbs—made them difficult to translate.

> The translation was not difficult except for translating the sayings/proverbs that the women used. I also faced some difficulties in selecting some words which are typically Creole and the correct equivalent in English was not evident. (Jorge)

In addition to these kinds of problems, translation was also difficult because of the need to interpret the words in the context of the culture. For example, Adam asked his participant, Pauline, how she would feel if her partner began a relationship with another woman. Pauline used colorful sayings to answer the question. Adam further explains what she meant and how it addresses the question he asks. She replied to Adam's question:

> I can't do anything about it except I just have to tighten my stomach with a tie and bear it. (Pauline)

In his interview notes, Adam explained her answer,

> When relatives die, women tie a cloth around their stomachs to cope with the pain and sorrow. In this particular case, the woman is saying that if her husband began a relationship with another man, it would be an enduring pain that she would have to cope with.

To the outsider, her comments convey the idea that her partner's behavior would be significant, but Adam's explanation of the saying she uses tells us just how significant. Her pain would be as great as if her partner had died.

Insiders or Outsiders?

Students experienced these problems in data collection partly because they were outsiders, in a sense. Some had family or work ties with their participants, but as researchers they played the role of outsider during the interview. They assured the women participants that their interviews would be completely anonymous, but participants may have feared the interviewer would pass on their information. Cape Verde is a small country where privacy is difficult to maintain. Gossiping between neighbors, quarreling, and especially talk of wife/husband affairs are a common pastime. At the same time, the reputation of being a "good woman" who does not engage in extramarital relationships and does not have a man cheating on her is important to Cape Verdean women. Participants, therefore, may not have wanted to disclose information they feared would be passed onto others, or might hurt their reputation (Solomon, 1992).

But all the difficulties the students faced did not come from their being outsiders (Baca Zinn, 2001; Zavella, 1996; Maykovich, 1977). They also had problems with their interviews because of their insider status. For example, perhaps because the students were insiders and should have "known better," they were put in the awkward position of inquiring about topics that were sensitive in their culture (Platt, 1981; Emerson, 2001). Furthermore, students experienced problems as insiders (to the community but not to the research team) because they did not control the conceptualization of the research or the initial interview questions. They challenged this situation with Carter as an outsider, by criticizing some of the interview questions and we ended up altering some of them. In addition, during the interviews, students made choices to probe in some cases and drop a question in others when they felt the inquiry was becoming too invasive (Naples, 1996).

Despite these difficulties, because the students shared membership with the group being studied, they held special insight into the meanings of the interviews with the women. These insights were more valid and accurate than if we, as complete outsiders, had collected data alone (Cannon et al., 1988; Merton, 1972; Baca Zinn, 1979).

The multiple insider-outsider relationships in this research show the complexity of that concept (Naples, 1996). In research by others who have explored this problem, the insider-outsider relationship is often conceptualized as that between researcher and researched. This research revealed more than one insider-outsider relationship, however. It suggested the ways in which insiders like the students are simultaneously outsiders. In looking at the insider-outsider relationship between the students and the participants another complexity is revealed, this time in terms of what makes insiders or outsiders. Much of the discussion in the literature shows the importance of differences in social class, race ethnicity, and gender between scholars and participants. In our relationship with the students and to the participants, these variables were most salient. In the students' relationship to the participants, additional factors come into play as well.

In Cape Verde, place of birth and residence, more specifically north/south and city/country and language dialect are important social delineators. These marked the place of both the interviewers and the interviewees in Cape Verdean society and marked them as insiders, outsiders or both. For example, most Cape Verdeans live in the *fora*, the rural, isolated interior of the country (70% of the population). Many of the students conducting the interviews came from Praia and Mindelo, the two largest and most cosmopolitan cities of the country. Women participants from the *fora* were likely to feel intimidated by and hesitant toward their urban interviewer.

Language dialect was also likely a social delineator. While Creole of the north is said to be more "sophisticated" and "Portuguesed," Creole of the south tends to be more "traditional" and "African." Women participants from the south may have felt intimidated when talking to their interviewer from the north.

Although these particular variables are often not taken into consideration in research in the United States, the city/country residence and language dialects may also be important determinants in the interviewer/participant relationship and data. We might learn from this experience about the range of ways we need to be sensitive to outsider insider state. In Cape Verde they are definitely critical issues that can intervene between interviewer and participant and influence the validity of the information obtained. Although we might be aware of the abstract principal that social variables can be important in the research experience, exactly what those variables are can be quite different from one culture to another.

Insider/Outsider Debate

Contemporary ethnographers feel obligated to tell their readers who they are and why they write as they do, since all studies, perhaps especially ethnographic ones, are enormously influenced by the personalities, training, preconceptions or theoretical biases, previous experience, and goals of those conducting the study. In the 1970s and 1980s, social scientists began to examine more closely the role of ethnographer in relationship to that of the informant (DeVault, 1996; Zavella, 1989). This became known as the insider/outsider debate because it explored the differences between the work carried out by insiders (individuals who belong to group being studied) and outsiders (individuals who do not belong to group being studied). This perspective emerged from research conducted in minority communities by researchers from dominant communities or in communities in Southern nations by researchers from countries in the North. The primary concern was the distribution of power in the field. Researchers who considered the issue were concerned that the differences in power between the researchers and the researched in these cross-cultural studies would interfere with collection of valid data. Those who contemplated this issue, in addition, were even more concerned that such encounters were ethically flawed and that the data generated would serve to perpetuate the political inequality because it presented the world from the point of view of those with more power (Reinharz, 1992; Andre, 1979; Powers, 1986).

The insider/outsider debate questioned how the social characteristics of the researcher influenced the level of interaction and participation among the subjects of that research who do not share those characteristics (Cannon et al., 1988; Merton, 1972; Baca Zinn, 1979). A central weakness in research that brings together people of different social strata is the quality of data collected and the validity of the conceptualization of the research and the interpretation of findings (Blauner & Wellman, 1973; Ladner, 1973). Carrying out research cross culturally is in addition challenging because differences in language, interpretation, and cultural values add to the complexity of conducting any research. Cross-cultural problems occur when differences in thinking, assumptions, behavior, and values become muddled together and misunderstood. To prevent this, researchers must be vigilant about how their own culture has shaped their ways of thinking and has placed within them the values and assumptions that oversee their behavior.

This is not an easy task because it demands that researchers not view the thinking and behavior of those they are studying as *normal* or *abnormal* but simply the way things are (Stewart & Bennett, 1991). Real people, researchers included, however, cannot help but think that their way of doing things is the best or even the only way. When researchers allow their own culturally bound perspective to get in the way of asking questions and interpreting answers in ways that flow from the culture of those they are studying, their findings are not valid and miss the core goal of mandate of contemporary ethnography to "understand the way that group members interpret the flow of events in their lives" (Agar 1980, p. 194).

In addition, their findings may distort "reality" in a way that creates misunderstanding of those they are studying and perpetuates the inequality between themselves and their subjects. Those who have entered into this discussion on insiders and outsiders have been particularly concerned about the role of racial ethnic relations in social sciences, and the prevalence of white researchers in the construction of ideas about the experience of people from other racial ethnic groups (Wilson, 1974; Blauner & Wellman, 1973; Beoku-Betts, 1994). These questions have also been asked for researchers who cross not only racial ethnic lines within nations but between nations as well, including studies done by Americans in Cape Verde (Andrade, 2000).

Feminist methodologists have been especially interested in these kinds of issues and have given attention to the diverse perspectives of women with various class, racial, ethnic, and sexual statuses and the ways in which social status or political power of the ethnographer and informant influences what is said, what is heard, and how it is constructed (DeVault, 1995; Krieger, 1985; Oakley, 1981; Stacey, 1988; Naples, 1996). Although feminist methodologists are more critically analyzing their influence and role in shaping the research process, they have not yet created entirely satisfactory methods for conducting research that overcomes these problems and builds knowledge in a way that both provides clear, accurate, and valid information and empowers the participants involved (Maguire, 1987).

Asking Questions

In the appendices we have included the questions in our interviews that pertain to the data reported in this book. Appendices A, B, and C include the questions students answered about the current economic and political situation in Cape Verde, the role of language in Cape Verdean society and the dance form of *batuku*. Appendix D includes questions the student assistants asked of the women they interviewed about work, family, and personal relationships with men. Appendix E lists the questions asked on the open-ended questionnaires distributed at the women's rights conference that were filled out by the women who participated in the conference.

We developed the questions through a process which involved proposing questions to each other and to our student interviewers and then revising them at each stage. We also altered questions, as we described earlier in this chapter when student interviewers told us that the questions were not effective. The initial questions emerged from Carter's informal observations before she began to methodically collect data.

In all of the questioning, our goal was to get participants talking (or writing) about their experience. We asked the questions in an open-ended format and (mostly) by beginning with words such as how, what, and why in order to encourage them to provide fuller explanations of their experiences and their ideas.

Compared to close-ended interviews (the participant chooses from a pre-determined list of answers like a multiple choice exam), open-ended interviews allowed the women participants to talk freely about their work situation, relationships with family, children, men and the community. As a tool of ethnography, open-ended interviews allowed the study to evolve, in response to what we were learning, as the research developed. As opposed to research guided by a hypothesis to be tested, open-ended interviews gave us flexibility to think about and address new ideas as they came up during research (Creswell, 2003).

After the data were gathered, transcribed, and translated we used a thematic analysis to examine the texts. The first set of data was the written responses given by the students who described the political and economic context of life in contemporary Cape Verde. The students described their perceptions of Cape Verde since the first multiparty elections in 1991 which we identified as a significant date marking a transition to a more open government and one that claimed to be acting more democratically, representing all Cape Verde citizens (Chabal, 1996). The students asserted that since 1991, the economic situation has improved for a few, mostly wealthier Cape Verdeans, but has deteriorated for the majority as the rich became richer and the poor poorer. The students believed that the government is responsible for these developments. They also believed that the government is capable and responsible for turning the situation around. But important changes need to be made in the government's actions. The students suggested that the government focus its attention on ending poverty, developing a strong independent nation, and ridding itself of corrupt politicians. These data are reported in chapter 4. Students also provided written responses on the importance of language and dance and this is reported in chapters 7 and 8. A second set of data came from surveys filled out by rural women attending a women's rights conference describe various challenges in their lives regarding the issue of violence against women. The significance of the problem of violence in women's lives is described in chapter 6.

The third set of data came from interviews the students conducted with women in their community. Two major themes emerged from these interviews. The first was the importance of gender inequality, especially in personal relationships with men in addition to the issue of economic survival. This is reported in chapter 5. The second major theme was forms of resistance to these difficulties which did not take the form of conventional political activities such as organizing for policy reform around violence against women as reported in chapter 6. Rather the more surprising forms that emerged in these interviews were *Batuku* and Creole proverbs as responses to the social circumstances of living in Cape Verde. These are the subjects of chapter 7 and 8. These themes—survival, men, dance, and language—appear to be dominant in the practical everyday experience of the women. They emerged in a thematic analysis of the data (Parker, 2005).

Developing a Thematic Analysis

Holloway 1997, (p. 44) writes that thematic analysis proceeds through several steps:

1. Ordering and organizing collected material.
2. Rereading the data.
3. Breaking the material into manageable sections.
4. Identifying and highlighting meaningful phrases.
5. Building and comparing and contrasting categories.
6. Looking for consistent patterns of meanings.
7. Searching for relationships and grouping categories together.
8. Recognizing and describing patterns, themes, typologies.
9. Interpreting and searching for meaning.

We used all of these techniques. Furthermore we used them as a research team with two authors exchanging data and interpretations in addition to the reflections and comments from the student interviewers about how best to understand the interviews. This process allowed us to develop first codes and then themes.

Juliet Corbin and Anselm Strauss (2008) write that coding is the way in which analysis takes raw data and raises it to a conceptual level. By this they mean that researchers must process the raw data, the words and ideas the participants give to the interviews. In that processing, researchers transform the text of the interviews (the raw data) into a more distilled version which summarizes and reflects the data which they call codes.

Sometimes codes arise from the literature (Miles & Huberman, 1994). When researchers use this technique they search for codes they have been led to believe may be part of the interviews because of what they have learned about the issues from previous research. For example, research on teenagers that work for wages has found that boys who earn money feel closer to their parents than boys who do not, while girls who earn money feel more independent of their parents compared to those who do not (Steinberg et al., 1982). On the basis of this knowledge of the literature, a researcher might identify codes of independence and closeness to be searched for if he or she were to analyze interviews of teenaged employees.

Codes can also emerge directly from the data (Boyatzis, 1998). In this case, the codes attempt to capture what seems most salient, common, or significant in the interviews (Cheek, 2004). This is the approach we took. For example, we asked students about life since the transition to a more open government in 1991. When we coded their answers line by line, we saw issues such as poverty, wealth, and corruption emerge as most common, salient, and significant.

Coding Data

After the interviews were conducted, we used a software program, Ethnograph 5.0 to help us sort, organize, and analyze the data. Our objective was to first break the material down into small parts and then to put it back together again in clusters that summarized the ideas and issues within the interviews. Our goal throughout the process was to move from the most concrete form of the data—the words the participants spoke or wrote in response to our questions—to more and more abstract ideas that captured the meaning of responses. We broke each interview into words first and then we developed categories to summarize the ideas in the words. Next we identified more abstract concepts that pulled together the categories. Kathy Charmaz (2006) explains this procedure, "Qualitative codes take segments of the data apart, names them in concise terms, and propose an analytic handle to develop abstract ideas for interpreting each segment of data" (p. 45).

The process we used to accomplish this was to first enter hundreds of pages of interview text into the Ethnograph program. We then read through the interviews many times looking for codes (which are sometimes also called concepts) that we felt captured the most important messages in the interviews.

In order to find the codes, we employed a specific protocol suggested by Ian Parker (2004), for both the discourse analysis and the thematic analysis. Parker's (2004) method of critical discourse analysis involves the identification of "objects" in the text through the primary identification of nouns used. According to Parker, nouns in a text mark the objects of analysis and by identifying them a researcher can "treat the text itself as the object of study rather than what it seems to 'refer' to." (Parker, 2004) Through this entry point of noun identification we can begin to identify themes in our data set of interviews. In our analysis, we read the interviews and created a list of all of the nouns. By concentrating on nouns as an identifiable part of text Parker (2004) maintains it is possible to engage analytically rather than simply "reading" the text.

Although the analysis was initiated through noun identification, nouns were not counted, as we were not concerned with a quantitative approach dealing with the frequency of particular nouns. Rather, this method was used to illuminate the character of the particular themes in the interviews. As we read the texts we noted if specific nouns, for example, infidelity, corruption, or poverty occurred often. Since our analysis was qualitative and not quantitative, however, we were not concerned with the number of times a word appeared but only whether it appeared across and throughout the interviews.

We then further grouped the concepts into a smaller number of themes, what Charmaz (2006) refers to as "handles." As we processed the data in these two steps, certain theoretical ideas struck us. We wrote these ideas down as short notes to ourselves, or memos. These memos helped us later to develop an even more abstract sense of the meaning of the interviews. For example in the interviews, women describe their relationships with men. Many of the women explain the relationships are difficult and unsatisfying. We assigned a code word of "difficult relationships." Women further explained the difficult relationships are mostly because of their partner's non-monogamous behavior. We then assigned a code word for "infidelity." Women have different ways of coping with unfaithful relationships and this became another way we grouped the data: "acceptance," or "giving an ultimatum." All of these codes might be summarized into the theme of seeing men as the problem. And at an even more abstract level it might be called identifying a system of masculine hegemony as the root of their problems.

Analyzing data means listening closely to what participants say as well as paying attention to what they choose not to say, or what they have not deemed important or significant. For example, we found it interesting that women did not mention the large out-migration of men as being a part of or the source of the problem. We knew that this was an important piece of the social context of the difficult personal relationships about which the women spoke. Furthermore, we knew from our reading that the out-migration was itself an aspect of the global economy which distributes money and jobs unevenly making it inevitable that people will have to leave economically depressed areas in the Global South to seek work in those nations in the Global North that are more prosperous and offer greater opportunities. This parallel (or encompassing) power structure that shapes gender relations remains invisible to the women. An economic power structure that requires that working people migrate in order to make a living and that makes life difficult for those who remain in Cape Verde plays an equally important but largely invisible role in shaping gender relations and creating distorted, unequal and unhappy relationships between women and men in Cape Verde.

As you will see in the chapter 5, which reports these data, we then put these two themes together: seeing men as a problem at the same time the larger political and economic context remains invisible. This contrast between what is visible and what is not can be further abstracted to reveal the collision of two stratification systems: gender inequality and international inequality (some nations are better able than others to provide those who live within their borders with opportunities to find a job and make a livable wage). And, this collision can be addressed by looking at the theoretical framework of intersectionality, which recognizes the many threads that hold us in place and the ways they intersect with one another (Collins, 2000). This transition from the most concrete raw data of codes such as infidelity and invisible issues such as men's migration to the most abstract level of the theoretical framework of intersectionality sums up the analysis of data throughout the text.

We used this procedure of developing a thematic analysis to analyze the data from the students in chapters 4 and 7; from the women who participated in the women's conference in chapter 6; and from the interviews the students conducted with the women about work and family that addressed issues regarding men and survival in chapter 5.

For the data we obtained in the interviews with the women and from the students' responses to questionnaires regarding the use of Creole, we used another related form of analysis, critical discourse analysis which is reported in chapter 8.

Critical Discourse Analysis

Critical discourse analysis examines the way people communicate with each other. These communications are referred to as texts. They can be found in a range of material including books, poetry, advertisements, conversations, or interviews. The technique has been most closely associated with people who study literature and language. But it is increasingly used in a broader way by scholars from many disciplines who investigate the ways people think, speak, experience, and express themselves in social relationships (Johnstone, 2002).

Critical discourse analysis, like critical thematic analysis begins with data collected through qualitative methods. It also shares with critical thematic analysis the commitment to explicating the underlying political relationships that frame the ways that people communicate with one another. In addition, it shares with critical thematic analysis the desire to study social problems in order to take action to solve them. "Critical discourse analysis has an explicit political agenda. It is 'engaged and committed'. It 'intervenes on the side of dominated and oppressed groups and against dominating groups' and 'openly declares the emancipator interests that motivate it'" (Fairclough & Wodak 1997, pp. 258, 259 quoted in Jones 2004, p. 98). Critical discourse analysts' focus on politics and their commitment to act politically is identical to the stance of critical ethnographers. "The critical objective [of critical discourse studies] is not only to identify and analyze the roots of social problems, but also to discern feasible ways of alleviating or resolving them" (Fairclough et al., 2004, p. 1).

Critical discourse analysis, however, differs from a critical thematic analysis in that it looks beneath the words to investigate the underlying, unintentional, "hidden agenda" of the ways people think about and talk about social issues. A critical thematic analysis explores the relationships between the interview data and the social frameworks that comprise the power structures. Critical discourse analysis questions the interview data itself, assuming that the words people use to answer the questions (or to ask them) are themselves part of those relationships of power. Norman Fairclough (1992) explains, "discourse is a practice not just of representing the world but of signifying the world, constituting and constructing the world in meaning" (p. 64). Further, he and his colleagues explain in their introductory article in the launch of the journal *Critical Discourse Studies* in 2004, "people not only act and organize in particular ways, they also represent their ways of acting and organizing, and produce imaginary projections of new or alternative ways, in particular discourses" (Fairclough et al., 2004, 1). Sometimes these projections serve to maintain the status quo and sometimes they challenge it. Critical discourse analysis studies the way that power, dominance, and inequality are enacted, reproduced, and resisted by text and talk (Van Dijk et al., 2003).

In the chapter on Creole, the interviewees provide the link between their use of language and the political and economic history of Cape Verde. Creole has been a means of communication shunned and degraded by the powerful, but preserved and celebrated by the population who speak the language. Just speaking the language itself is an act of defiance and solidarity with others who speak Creole. Creole has also been put to good use by the women in Cape Verde in explicitly expressing solidarity, providing solace, and documenting criticism of the power structures.

Our analysis of the use of Creole in Cape Verde uses the tool of critical discourse analysis. We explore not only what the women say when they describe troublesome relationships with men, we also explore *how* they say it. Discourse analysis:

> challenges us to move from seeing language as abstract to seeing our words as having meaning in a particular historical, social and political [context]. Even more significant, our words (written or oral) are used to convey a broad sense of meanings and the meaning we convey with those words is identified by our immediate social, political and historical [contexts]. Our words are never neutral. (Fiske, 1994)

We listened to the women's words, the discourse captured in the interviews, and we tried to hear the underlying meaning. For example, when Adam was interviewing Pauline, he asks her how she would feel if her partner began a relationship with another woman. To answer the question, Pauline describes tying her stomach with a tie and bearing it. We then analyze her discourse and see the meaning it has within the social context of Cape Verde. As Adam explained, her comment refers to a symbolic way that women signify and bear their grief over a deceased loved one. Pauline said if her partner began a relationship with another woman, her pain would be as great if someone had died.

But we also look at *how* she has conveyed this idea. Instead of explaining it straightforwardly, she uses a saying. This allows her to communicate her strong feelings in a manner that is less threatening. This manner of communication implies a power relationship in which Pauline does not have the authority to speak out in a way that overtly expresses the strength of her emotions. The manner of communication tells us about the power structures in which Pauline is embedded and the vulnerable position in which she finds herself. On the other hand, because she has been able to communicate her opinion, she is in a sense challenging that power structure and asserting her authority, her right to denounce infidelity and the exploitation of women by men in personal relationships. Critical discourse analysis helps us to understand Pauline's comment more fully and especially to understand the political context that shapes the experience and meaning of her description. In chapter 8, we develop this line of analysis looking at the direct and indirect meanings of the discourse in the interviews by using the tool of critical discourse analysis.

Reflections on Chapter 3 Research Methods

Every researcher faces an array of possibilities for finding ways to try to answer a research question. Our research question is what has been women's experience in the globalized economy of the twenty-first century? What problems have they faced and how have they responded to them? This chapter describes the process by which we decided which techniques to use to gather information and what we did to try to understand the data once we had collected it.

The chapter has defined terms such as qualitative, ethnography, critical, thematic, and discourse to show where the techniques we used to conduct our study fit into the literature on social science research methodologies. We have provided some historical information so that the reader can see where we place ourselves in the development of research methods. Furthermore we have described some of particular strengths of the methods for answering the questions about which we are most interested.

In addition to discussing these general questions of what a methodology is and what methods are best suited to our problem, we have also provided a detailed overview of how we went about gathering data and analyzing the information. This serves several purposes. First, it is essential for researchers to detail the specific methods used to gather and analyze data in order to be clear and transparent to him/herself and to the academic community as well as to the reader, to prove accountability, responsibility, and to assure others that ethical decisions and actions were taken.

Another reason to detail the methods used in research relates to the issue of objectivity. Researchers always have a point of view when beginning a study. The personal characteristics and objectives of the researcher impact their perceptions of the culture as well as the data collection. Paul Rabinow, author of *Reflections on Fieldwork in Morocco* writes,

> There is no privileged position, no absolute perspective, and no valid way to eliminate consciousness from our activities or those of others.... We can pretend that we are neutral scientists collecting unambiguous data and that the people we are studying are living amid various unconscious systems of determining forces of which they have no clue and to which only we have the key. But it is only pretense. (Rabinow 1977:151–152)

Complete separation of researchers from their studies and its results are impossible and influencing the data to some degree is inevitable. Researchers, however, should strive to be objective in the sense that they should try to present data in a manner that is not purposefully distorted and as fully aware of the sources of subjectivity as possible. Researchers, must, therefore, recognize their own point of views and biases. They must, furthermore, identify their position and let the reader decide about the validity of the study.

In this study we have identified ourselves as scholars who are concerned about social inequality and injustice and wish to explore our social world in order to determine ways to help make it a better place to live. We also, however, are committed to finding out what is really happening in our world as a first step to changing it. Therefore, we provide this methods chapter as a way to clarify for ourselves and others what we did and where we may have missed seeing or misinterpreted critical information. Delineating research methods as we have in this chapter is important because if someone questions our research they will be better able to see where we got our information and the logic behind our decisions at every step of the research process and most importantly the logic behind our conclusions.

Finally, describing the details of methodology also is useful to those who might be interested in conducting a similar study. Here they can see step by step what exactly we asked in our interviews, for example, and how we conducted our analysis. Just as we have benefited from studying the work of other scholars, we hope they can build on our work.

4

Cape Verde in a Global Economy

Rich people live in a big and nice house, have nice cars, and their children are well treated wherever they go. Meanwhile we have poor people who hardly have food.

—*(Natalina)*

As the twenty-first century opened, Cape Verde was among the poorer nations of the Global South. International economic institutions have recently changed Cape Verde's status from the very poorest category "least developed country" to the status of "lower medium developed country." With a gross national income of $596 million, which amounts to $1,310 per person per year, however, it is still considered among the poorer nations in the world (*Europa World Year Book*, 2003).

Agriculture, together with forestry and fishing are the major employers of the islands. Almost one quarter of the economically active population are working in these areas but the businesses are not thriving. Agriculture has been plagued by drought for decades and little land (12%) is suitable for cultivation; 80 percent of the food eaten in Cape Verde is imported.

Fishing is equally bad. Despite the fact that Cape Verde is located far off the Western coast of Africa in the middle of the Atlantic Ocean, fishing remains small because of a lack of modern equipment. For example, of 1,400 fishing boats in the nation, only around 40 percent are motorized, and only 64 of the 95 large vessels used for industrial fishing are fully operational. As a result, the fishing industry has declined because investments have not been made in boats and other equipment. To illustrate this downturn, the fish catch fell from 14,730 metric tons in 1981 to 10,821 in 2000 (*Europa World Year Book*, 2003).

Manufacturing is another important employer and source of GDP. But Cape Verde suffers from lack of resources, high prices, and poor location. Goods are made from raw materials that are expensive because they must be imported and Cape Verde is not located on the major shipping routes. These weaknesses in turn discourage investors from putting their money into Cape Verde (Johnson, 2004). Some economists have suggested the difficult situation might be remedied by transferring to a service based economy, particularly one centered on tourism (Afrol, 2006).

The service industry accounts for an increasing piece of the economy and most employed people currently work in a service job. Tourism is perceived as the area with the greatest potential to create jobs and boost the economy (*Europa World Book*, 2003; Scheuermaier, 2004). However, there are concerns that tourism could create a risk to the sea life; specifically loggerhead turtles and humpback whales (BBC News, 2008), and other vulnerable species including the blue whale and harbor porpoise (AnimalInfo, 2005). Furthermore, building the tourist industry will require a huge infusion of capital to build the resorts and infrastructure.

Whether tourism is the answer is yet to be seen. In the meantime, the World Bank reports that Cape Verde continues to be a debtor nation with a total external debt of $360 million at the end of 2002. The economy grew by an average of 6.1 percent every year from 1996 to 2001. At the same time, however, debt has risen and almost one quarter of the population remains unemployed and an additional 26 percent are underemployed. In addition, the poverty rate is increasing, up from 30 percent in 1989 to 37 percent in 2003 including 20 percent who are identified as extremely poor. 66 percent of women and 55 percent of men in Cape Verde are currently reported as not economically active and the country is ranked 38 in the Human Poverty Index by the United Nations (International Monetary Fund, 2005; United Nations, 2008; World Bank Development Report, 2006).

Besides being indebted to countries of the North, Cape Verde also is part of the global economy in other ways. It is among the largest recipients per capita of foreign aid in the world having received $325 for each Cape Verdean in 1999 (Scheuermaier, 2004). In addition, major industries are owned by nations of the North. In 1993, the government passed laws opening the economy to foreign investors. Between 1993 and 2000, $180 million were invested by corporations outside of Cape Verde, mostly from Portugal in manufacturing enterprises and more recently from Italy in the tourism industry

Table 4.1 Cape Verde: A small country with a long history of economic difficulty

- The land surface is 4,033 square kilometers (Solomon, 1992).
- The country survives on subsistence agriculture, animal husbandry, remittances from emigration, and aid from other countries (*Europa World Year Book*, 2003).
- From 1750 to 1990 a total of 60 years of famine were responsible for over 250,000 deaths. Drought continued in the 1990s and Cape Verde appealed for international food aid in 2002 to head off famine (BBC News, 2008).
- 85% of all products used in Cape Verde are imported (Scheuermaier, 2004).
- Only 12% of the land can be cultivated (CIA World Fact Book, 2008).
- The nation depends on remittances from the 700,000 Cape Verdean emigrants living in other nations (*Africa Research Bulletin*, 2006).
- Life expectancy is 67 years for men and 73 years for women (BBC News, 2008).
- Cape Verde is a "structurally food insecure country" and is prone to food shortages since it can only produce about 10 percent of its food requirements each year. Widespread malnutrition is a constant problem (Afrol, 2002).
- The nation carries $360 million dollars in external debt and per capita annual income is $1310 (*Europa World Year Book*, 2003).
- Cape Verde ranks 100 among 173 nations in the HDI (Human Development Index) and 82 among 146 nations in the GDI (Gender Development Index) (United Nations Development Program, 1999).

(Scheuermaier, 2004): 49 percent of the nation's banks, hotels, airlines, and shipping lines formally owned by the Cape Verdean government have been sold to foreign investors. Portuguese investors also own 40 percent of the state telecommunications company (Africa—South of the Sahara, 1997) and 51 percent of the state's electric company and water supplier, Elektra (Afrol, 2003a).

The dominance of foreigner investors in even the industries that supply the most basic needs, such as water, are a result of policies of privatization, a key element of neoliberal and Washington Consensus economic "reforms." And the IMF continues to push its privatization drive demanding that Cape Verde privatize its few remaining public enterprises, including the national airlines (TACV), ENAPOR (National Company for Port Administration), and ENACOL, the state's main oil supply company. In 2001, the state owned transport company, Transcor, was bought out by the privately owned transport company Moura, resulting in price increases for the consumer (World Bank, 2007; Afrol 2003a). Both the country's private sector business class and poor Cape Verdean households have been greatly harmed from this privatization rampage. The IMF claims that privatization is the best way of making businesses efficient and productive because it causes them to enter into a competitive market. Critics, however, point out that privatization often results in lower wages for employees and shoddier products for consumers as management seeks to reduce its costs and maximize its profits (Campaign for Labor Rights, 2003). Privatization also leads to job loss and price hikes. On one island, the increased cost of privatized energy has forced people who cannot afford the price hikes in electricity to return to traditional oil lamps (Afrol, 2003b).

Like many nations in the Global South, Cape Verde is going through a period of major economic challenges and change. Events are unfolding as a result of huge global forces. People are being swept increasingly into a global economy where decisions are made in the North and the benefits are not returned to people in the South (Mazur, 2000). The people in Cape Verde must live through these changes. It is apparent that, for them, globalization is not an abstract economic theory but a real day-to-day experience.

National Politics in Cape Verde

In 1991, Cape Verde had its first multiparty elections. Cape Verde won its independence from Portugal in 1974 and held its first election in 1975 (U.S. Department of State, 2008). Until 1991, however, one party, PAICV (African Party for the Independence of Cape Verde) was the only one to run candidates. After 1991, the electoral system became liberalized and other parties from a broad range of political viewpoints joined the PAICV, including the MPD (Movement for Democracy), PAIGC (African Party for the Independence of Guinea and Cape Verde), PCD (Democratic Convergence Party), PRD (Democratic Renewal Party), PSD (Social Democratic Party), PTS (Party of Labor and Solidarity), UCID (Independent and Democratic Cape Verdean Union) to run candidates. The PAIGC and the MPV has alternately won control of the national government since 1991.

When multiparty rule began, the government initiated a series of policy changes that promoted globalization as a solution to the economic difficulties the nation faced. Cape Verde has tied its fortune to investment from wealthy nations and has been willing to abide by the reforms those nations and international financial institutions require of them. BBC News reported in 2006, "to help boost the economy, the government has tried to lure foreign investors, especially in the tourism sector." The government has also sought closer economic ties with Europe and the United States. The United States has pledged to give Cape Verde $110 million as part of its Millennium

Challenge Account initiative, which rewards poor countries that show a commitment to "economic and government reform" (BBC News, 2006).

Since 1991 was such an important political date for Cape Verde we decided to use it as a marking point for our research. We wanted to see how the new government policies had affected people from different social classes since the opening up of elections in 1991. We asked students questions, such as, how have things changed for rich people and poor people since the new government began in 1991? What is your perception of the government's role since 1991? What should the government do to help solve the problems that Cape Verde faces? (For a complete list of questions, see appendix A.)

Rich Are Richer and the Poor Are Poorer

In 2002, during the time period we were conducting our study, another group of researchers (Ames et al., 2003) conducted the Afrobarometer survey in Cape Verde on a sample of 1289 individuals on the four major islands. The survey utilized essentially the same Afrobarometer Round 2 questionnaire implemented in 12 other African nations (Botswana, Ghana, Lesotho, Malawi, Mali, Namibia, Nigeria, South Africa, Tanzania, Uganda, Zambia, Zimbabwe). This report traces the relationship of historical, economic, and demographic factors on Cape Verdean evaluations of democracy and market reforms (Ames et al., 2003). The survey found that 56 percent of the population believes that Cape Verdeans' standard of living had not improved since 1991 when the government reforms began and globalization became a central force in their nation. 68 percent said they agreed with the statement "government policies since that date had hurt most people." Many respondents also felt the economy was not doing well. 40 percent rated the current economy as bad, 45 percent said it was neither good nor bad, while only 10 percent said it was good. The students we surveyed for our research had similar opinions. Overall students felt that either only rich people had benefited from changes in the economy in the past decade or that poor people had benefited only in small or unstable ways. Some of the students felt a combination of changes had occurred since multiparty elections. But most students believed that the rich were richer and the poor poorer or at least that a gap remained. In the general survey conducted by Afrobarometer in 2002, the gap between the rich and poor was named as the second most important problem facing Cape Verde. They identified unemployment as the most important problem (Ames et al., 2003).

For the students we interviewed, however, the gap was most significant. Some students, like Samira (all of the student's names are pseudonyms) focused their attention on the rich in their reflections on the gap between rich and poor. These students emphasized the ways in which they believe the rich have improved their own situation since 1991 and now have better opportunities, including better access to health care and education, and higher status:

> The rich became richer; they improved their opportunities in life. (Samira)

The student's perception of these improvements did not stop at more money or even the things money can buy. They also mentioned that along with improved material conditions, the rich enhanced their social situation and were better able to take political leadership, interact with others, and even enrich their leisure time. Natalina listed several ways rich people's lives had improved since 1991 and the globalization of the Cape Verdean economy:

> I think that the rich became more and more rich, because they are invited to engage in [political] parties, lead groups, and receive people. Their children have great education, get good jobs, and have good medicine and health care. (Natalina)

Furthermore, students noted that since 1991, rich people have secured their position more strongly. They described the increased control the wealthy now have over their social environment which in turn means they are now better able to secure their future. For example, Dani explained how wealth translates to power when rich people are able to influence prices in ways that benefit them. He asserted that having wealth gives people power in the economy:

> I think that rich people have become richer and richer because the free market economy has allowed them to use the prices they think is good for them. (Dani)

Dani asserts that the rich, who are business owners, are not restrained by the government in their efforts to keep labor costs down and prices up, thereby improving profitability. The IMF has promoted the idea of privatization and according to Dani, this is a useful way to develop the economy but it has only benefited wealthy business owners who have increased their profit margin.

And the Poor Get Poorer

Other students focused their attention on the poor and argued that the poor became even poorer since 1991 as wages stagnated and prices rose. This is the flip side of the trends Dani identifies as benefiting the rich. The rich getting richer is tied to the poor getting poorer although students made these observations as distinct from one another. The students mentioned they had heard that the economy had changed for the better for everyone but they are not persuaded by those who claim the situation is improving. Relying on their own observations, they maintained that the situation is not getting better for poor people:

> Some people say that it became easier for people to get a job, to have food, but I don't think so because the situation is not getting better. The poor people are staying behind. (Emilia)

They believe that the economic system does not serve the interests of the poor. According to Dani, the poor must struggle for their material needs and they have no chance of exerting any power that could alter the system in a way that would improve their access to goods and services they need to survive. He explains the links among wages, prices, poverty, wealth, and power:

> Since the prices of things are in accordance with the interest of those who are rich, these people won't stop rising the prices whenever they get a chance to do so. In a situation like that, the poor will have to spend more and more money. And because the wages are not high and because the increases of prices do not keep pace with the rise of wages, there is a tendency for the poor to become poorer. (Dani)

Students were asked to think of an example that showed how the gap between rich and poor had grown larger since 1991. Most described wealthy people as living a good life enjoying life, liberty, and the pursuit of happiness, while poor people are searching for the barest necessities:

> Rich people live in modern apartments or houses or in fashionable, expensive private homes or they drive modern, expensive cars, their

> children vacation in USA, France, Italy, or Brazil every year. But poor people live in ghettos without even food three times a day. (Fernando)

Floors were a described as a particularly important marker of differences between rich and poor. They indicate the pervasiveness of the social class distinctions, *even* floors:

> Even the floors rich and poor walk on are different. (Emilia)
>
> Only rich people live in Palmarejo [a wealthy neighborhood], because the floor [putting tile on one's floor] is very expensive so the poor can't buy it. (Cabral)

Modest Improvement or Sharp Decline since 1991

In the Afrobarometer survey, availability of goods was the improvement most often identified by the subjects when they were asked what was better now. 67 percent of the respondents said that since 1991, Cape Verdeans had better access to consumer goods. The students who felt that life in Cape Verde has improved since 1991 for the poor as well as the rich also told us that more goods were available. In addition, the students mentioned that at least potentially there were more resources with which to try to solve important social problems.

Emanuel, for example, said that he believed that Cape Verde now had more tools with which to fight poverty. He felt that other countries were helping more but he couldn't say what exact benefits that had for low income Cape Verdeans. His comment suggested that the *potential* for improvement has improved but real benefits are still elusive:

> Since 1991, poor people have found a better life, because of the political changes. Nowadays we have some ways in trying to fight poverty in Cape Verde, for example we receive help from other countries. (Emanuel)

Cabral noted some specific improvements in access to consumer goods, such as a more diverse diet. This change is undoubtedly significant in a country where a diet of rice and fish can become monotonous and the lack of fresh fruits and vegetables unhealthy. But the specific changes he noted are small. Perhaps Cape Verdeans are moving in the right direction but these are certainly only modest gains:

> Now everyone can eat apples and drink Coca-Cola, because they are cheaper than before 1991. Now we don't have to depend only on rice, corn, and sugar, because we are more liberal. (Cabral)

The gains, however, may be noteworthy in terms of the symbolism of the food. In a nation with a history of food production for export to wealthier nations and the rich within the nation, access to foods that are not necessarily healthy (Coca-Cola) or those that may appear to be relatively inconsequential (apples) may reveal profound meaning in social relationships. If the dockworkers that unloaded the crates of soda did not make enough money to purchase the drinks, and now they can, the change is filled with meaning beyond its superficial appearance.

Besides improvements in access to goods and services, some students spoke of more abstract advances in civil rights and educational opportunities. In the Afrobarometer survey respondents identified freedom of speech (80%), freedom of association (79%), and freedom of electoral choice (81%) as having improved since 1991 (Ames et al., 2003). The students agreed and noted some of the same gains.

> Nowadays, people's lives have become better. People were opened up to the world. Now we have freedom of speech for a better life. (Emanuel)

> We won the freedom of speech. One thing that has been helping us is the education of our children. (Natalina)

Like Natalina, Fernando mentioned the improvement in education since 1991. He said that the general population has greater access to education and some people have even been able to obtain higher education and study abroad. He notes, furthermore, that their opportunities to gain an education in other nations allowed them to "bring some wealth," that is, it allowed them to find good jobs overseas and to send some of their earnings home to families in Cape Verde:

> Poor people changed in many ways because before there were few schools in 1991. Nowadays, the government built more schools on all the islands. Poor people began to study and to reach the University level. Some began to travel abroad and to bring some wealth. (Fernando)

Miranda, however, suggested that although this is an important improvement, it was short lived and has disappeared in recent years. She stated that in the 1990s steps were taken to improve people's educational opportunities as well as health care. But the new century has brought further changes that have halted progress in these areas:

> During the period from 1991 [to] 2001, there was a boom and things improved for everybody. In 2001, we had elections and the new party in power has increased taxes and everything. So now, life is more difficult. During the first ten years, people's lives improved because lots of

> schools and hospitals were built. After 2001, there is almost no foreign investment and the economy is in bad shape. (Miranda)

The students as well as the participants in the Afrobarometer survey identified a number of improvements they felt had been made since 1991. When the students described the gains, however, many of them were relatively small or sporadic and often had evaporated or were only potentials or symbolic.

What Role Has the Government Played since 1991?

Students described why they thought the changes in the economy have taken place since 1991. They agreed the government was a key actor in creating past and present economic circumstances, both good and bad. They were generally critical of government officials maintaining that the policies the government created often exacerbated problems rather than solved them.

In the survey by Afrobarometer, only 33 percent of the subjects said they were satisfied with their democracy. This number is lower than any other of the 12 African nations included in the study, except Zimbabwe, where only 18 percent of the respondents said they were happy with their democracy. While Cape Verdeans believe democracy has been established in their nation and they believe that democracy is a good foundation for a government, 41 percent say that Cape Verde is a democracy with major problems and another 36 percent say it is a democracy with at least minor problems. Respondents were asked about how they thought the government was handling specific challenges. The proportion who said the government was doing well or very well in regard to specific social issues was as follows: combating HIV/AIDS (56%), addressing educational needs (54%), improving basic health services (50%), delivering household water (45%), ensuring everyone has enough to eat (34%), managing the economy (33%), resolving conflicts between communities (33%), keeping prices stable (29%), reducing crime (29%), fighting corruption in government (24%), creating jobs (23%), and narrowing gaps between rich and poor (23%) (Ames et al., 2003).

In our study, students who were critical of governmental programs said that sometimes the policies the government implemented were intended to help improve the economy to benefit everyone but ended up primarily benefiting the rich. The students also suggested that sometimes the policies were a result of explicit corruption by officials who promoted programs intending to ensure that only they would benefit by increasing their wealth and keeping themselves in power.

Edelton was among those who criticized the policies implemented by the government which he believed were ineffective or in fact intensified the problems Cape Verdeans face. He said that privatization policies were useful for the rich but they ultimately increased the gap between rich and poor. He believed that the economic policies of the new government were designed to help wealthy investors and ignored the plight of the poor:

> The rich became richer because the new political system allowed more private investments and gave them the chance to choose what they wanted to invest in. (Edelton)

At the same time, the government's manipulation of the economy has resulted in increased poverty as Miranda describes:

> There is a program for poverty alleviation and according to its statistical data the number of poor, which was 30% of the population, is now 36%. Most of them are women and live in the urban areas. (Miranda)

Some suggested that the government was in fact working against not only the poor but also the middle classes by raising taxes and by creating economic policies that increased prices and did nothing to reduce unemployment. Emilia describes the way that inflation and unemployment resulting from government policies has hurt the middle class as well as the poor:

> Everything is becoming more expensive, there are not enough jobs for everybody, and we have to pay more rent in housing. Water and electricity are more expensive too. (Emilia)

All of these observations were not focused on corruption. Some described the actions of the government as intended to maintain the status quo by ignoring the poor and serving the needs of the rich, but they did not necessarily identify these activities as illegal and corrupt. A second set of concerns centered more clearly on corruption. In the list above relating the findings of Afrobarometer, only 24 percent of the respondents felt that the government was effective in stopping corruption. Furthermore, 25 percent said that the current government, compared to the previous one was more or much more corrupt (Ames et al., 2003).

Fernando, agreed with the assessment of the 25 percent in the Afrobarometer survey. He explicitly described what he perceived to be thievery among political leaders. In particular he saw privatization as a program to take from the general population and give to the few:

> Since 1991, those politicians who came to power, some of them belong to former dictatorship parties. They had some money before, and they became even wealthier, because they stole the country's money by privatizing government companies and now they have most of the shares. (Fernando)

Miranda agreed that corruption is a problem. She described the specifics of the transfer of wealth from the general population to the politicians. She believed the tax system created a funnel for money to be taken from citizens and given to politicians for their personal gain. She, however, was still hopeful that democracy could prevail and suggested that this corruption can and will be stopped at the voting booth in the next elections:

> This government must lose the next elections (2006) because they have no economic policy. Their economic policy consists of increasing taxes, water, electricity, and phone charges to pay their bills. The burden is placed on the people. However, they, the Ministries, continue traveling all the time, changing their official car, moving to fancy houses, all of it paid with the official state budget. (Miranda)

Another Vision of Globalization

> *[The government] should give jobs for everyone, so the life of many people would change.*
>
> —*(Maria)*

Students were asked what the government should do to help solve the problems in Cape Verde. Some were pessimistic about the possibilities because they did not believe the government was committed to addressing the needs of the people. Dani articulated this caution:

> I think that the government is not interested in solving the real problem of the people. The poverty will continue, I believe. (Dani)

Nevertheless, students were creative and articulate when it came to describing specific ways they believed the situation could be improved through government policy if the government were to act on behalf of the public. They had many ideas about policies that could make a globalized economy in Cape Verde one that served the people.

They proposed that the government:

- Focus on poverty
- Develop a strong independent nation
- Get rid of corrupt politicians.

Focus on Poverty

The students asserted that the government must begin from the premise of eliminating poverty. Rather than creating policies that addressed other issues with poverty reduction as an afterthought, they suggested that poverty be the focus and met head on. They describe the government as an institution that could, together with private organizations such as NGOs, organize public finances and create programs that could develop the country, educate the population, and provide housing:

> I think the government should try to give opportunities to the poor people so that they can improve their conditions. (Alcides)

Emanuel took this position by describing what a good government would do in light of the growing numbers of poor people:

> I think a good government would help to diminish the poverty problem because population has increased in Cape Verde. So we have to have good managing of our resources, and try to save money, and do things that really contribute to the development of our country, like smart investment. (Emanuel)

Samira provided a specific policy he believed would address poverty and emphasize the need to put poverty front and center:

> Build more houses for poor people. Don't use poverty problem-money [funds earmarked for welfare programs] to solve other problems. (Samira)

Deusa and Luis also offer a specific plan to address poverty and in doing so improve the lives of many people:

> Join hands with NGO's and Volunteers in order to fight/solve the poverty problem. (Deusa)

> They must create more jobs; encourage the young people to study more and more. (Luis)

Cape Verde as an Independent Nation

Students also believed that Cape Verde should act independently. They did not speak of isolation but of developing economic plans that created interdependence among the different islands of the nation and creating a plan of action to interact with other nations while keeping the interest of Cape Verde at its heart rather than maintaining the dependent relationship between Cape Verde and rich nations:

> To solve poverty, the government should give equal rights and opportunities to all the islands in all matters. They should do something to create wealth and not waiting for and expecting only foreign aid. (Fernando)

In addition to suggesting that national integration and equality are important to building a strong nation, they also maintained that Cape Verdeans needed to be encouraged to develop their economy and educated about the best ways to go about that:

> The government should educate Cape Verdeans, to recognize that they don't have to wait for the government to employ them, but they can produce anything and employ others. To create a good environment that makes possible competition in the global world. To strengthen the bank loan as a way to establish companies, factories, etc. (Cabral)

Ending a Corrupt Government

Students identified a number of ways the government could at least theoretically address the needs of the people, especially the poor people. However, they also were not convinced these would be easy to implement because they believed the current government is a dishonest and fraudulent institution, and before the government could help solve problems, it must first organize itself in an honest way:

> The government should act as human beings and stop telling lies and stop stealing our humanity, dignity, and money. (Manuela)
>
> They have to stop stealing the government money because they did that before and that was a shame. (Fernando)
>
> Stop the corruption and make politicians, which are concerned with equal rights for everyone. (Luis)

The students believed that there is hope in democracy to implement all of these proposals but so far that hope has remained unfulfilled:

> Give everyone the same opportunity in life. In democracy it is said that everyone has the same opportunity but that isn't true. (Edelton)

A Critique of Current Globalization

In the late 1980s, Western leaders met to develop the idea of an economically and politically globalized world. Leaders developed this plan through stateless organizations such as the World Bank, the WTO, and the IMF. These organizations would include new globalized capital markets and a new world trading system that supposedly would lead to a well-organized world capitalist system and thereby reduce the gap between the world's rich and poor. Along with a renewed United Nations organization, the political aspect of the plan included a democratic model bringing freedom to formerly oppressed people around the world. The World Bank and WTO regarded globalization and a globalized economy as a great economic and political movement that would benefit everyone, including the Global South, as it spread the economic and social organization of the wealthiest and most powerful nations across the world (Callaghy, 1993).

Has this plan succeeded, transforming the world into some type of multinational utopia, or at least moving it in that direction, by creating a global economy with free trade and closing the gap between the world's rich and poor? According to many scholars (Rai, 2002; Ault & Sandberg, 2000; Acosta-Belén & Bose, 1995) as well as participants interviewed for this research, globalization has not achieved these goals and instead has hurt people in the Global South.

Participants of this research were generally critical of the kinds of policies implemented by Cape Verdean government officials since 1991. They argue that the government's policies often intensified problems rather than solved them. Sometimes these policies were intended to help improve the economy to benefit everyone but ended up primarily benefiting the rich.

According to the students who offered their views in our research, policies were designed to help wealthy investors and ignored the plight of the poor. The government's manipulation of the economy resulted in increased poverty. Other students described the ways in which privatization policies, for example, were useful for the rich but they increased the gap between rich and poor.

Privatization has been successful in some countries, according to economists. In Cape Verde, however, huge sections of the economy are owned by outside investors, and these investors do not reinvest their profits in Cape Verde. When the stockholders profit from their business, the money they make will not benefit Cape Verdeans or communities in Cape Verde. This chapter suggests privatization and IMF policies are not effective strategies for ending poverty in Cape Verde.

Students (as well as the general population represented in the Afrobarometer [Ames et al., 2003]) generally agreed with scholars who argue that globalization benefits only those at the top who live in the wealthy nations of the world, or those who comprise the ruling elite in poor nations, while the common person living in poorer nations sees nothing but economic hardship. Critical scholars assert that globalization makes life more and more difficult for poor people. Even Joseph Stiglitz, former chief economist of the World Bank, strongly criticizes globalization policies. He argues that free trade agreements mainly benefit the rich, and privatization has been disastrous in many countries. He states unless the rules of global capitalism are changed, the gap between rich and poor will continue while social and economic conditions will fall (Shah, 2001). Students also suggested that the rules of global capitalism should be changed.

Students asserted that globalization could work in Cape Verde but only with major reforms and an alternative vision. They first argued poverty should be eradicated. They didn't suggest that poverty should be simply reduced but rather the government should focus 100 percent on poverty elimination and devise policies to that end. They argued that together with NGOs, the government could create programs to develop the country, educate the population, and provide housing (Angen-Urdinola & Wodon, 2008). Cape Verde should act independently, according to the participants. The government could develop economic plans that created interdependence among the different islands of the nation and create a plan of action for an independent Cape Verde, rather than maintaining the dependent relationship between Cape Verde and rich nations. They also asserted that Cape Verdeans need to be encouraged to develop their economy and to be educated about the best ways to go about that. Nevertheless they point out that before the government can solve problems; it must first organize itself in an honest way.

The view of globalization in general offered by scholars and the specific situation of globalization in Cape Verde described by students provide a picture of great economic difficulty. They also suggest that the government could play a constructive role but instead has been mired in corrupt or inept practices that have done little to build the Cape Verdean economy in a way that benefits all of its citizens. This discussion sets the stage for looking at the lives of women in Cape Verde.

Reflections on Chapter 4 Research Methods

Chapter 4 begins to present the primary data we collected in our research. The information in this chapter comes from the written interviews we conducted with students in the first step of our data collection. The questions we asked, which focused on the economic and political context of a nation in the Global South and the perceptions Cape Verdeans have of their political and economic place in the world, were open-ended. We designed the list of questions in a way that we hoped would elicit their opinions with as little interference from us as possible. We chose the topics, but the answers the students gave used their own language and included the issues within those topics they felt were most salient. The qualitative character of the data collection allowed us to come as close as we could to the perceptions and experience of a group of Cape Verdeans.

We also have included in this chapter some data from a large survey conducted on a representative sample at about the same time as our research and focused on similar topics. Although we did not know of the survey until after we had begun writing this book, both our interviews and the survey conducted by Afrobarometer asked about the same general questions regarding the economic difficulties and successes in Cape Verde and the role of the government in addressing those issues (Ames et al., 2003). Both our study and theirs also ask about some of the specific issues that emerged in our data such as the gap between rich and poor, the continuing significance of poverty, and the problems of corruption and ineffectiveness in the government. We reported their study along with our own in order to provide a deeper look at the issues and to show the ways the data both dovetail and contrast with one another.

The juxtaposition of the quantitative data from the survey and our own qualitative data strengthens our understanding of the issues. The quantitative data provides us with a more powerful indicator of what people, in general, in Cape Verde believe about their nation's political and economic situation. The larger sample and its randomness mean that we can more assuredly assert that the ideas are representative. Our own data come from a small sample and we cannot confidently argue that the answers our participants provided reflect the nation as a whole. However, when we see our data along side the survey data, it appears that our respondents do have views that are fairly consistent with the range of views in the general population (as measured by the survey).

On the other hand, the qualitative data provided in our own interviews provides better clarity on the issues. We were able to get beyond short phrases and yes/no, higher/lower answers to find exactly what people think are the connections among poverty, globalization, and government or how exactly people experience the gap between the rich and poor. We were better able to see the meaning beneath the opinions and the emotions expressed in the discussion of the issues in our interview data.

Before reviewing the primary data we collected in the interviews in this chapter, we began by providing information about the political and economic context of the interviews focusing on contemporary Cape Verde. Already in this book in chapters 1 and 2 we have described the political and economic context, but in chapter 1 our focus was more broadly on the Global South, of which Cape Verde is a part. In chapter 2, we provided mostly historical information on the development of the political and economic character of Cape Verde. This chapter has picked up those threads and brought us forward in history and more closely focused on Cape Verde with descriptions of the specific nature of the economy and the details of the political organization, especially since 1991.

1991 is identified as an important turning point in the questions we used in the data collection. This chapter has explained how the decision to mark 1991 as a watershed was reached.

The role this chapter plays in the book as a whole is to polish and focus our understanding of the political and economic context of life for women in Cape Verde. The chapter has not provided us with much sense of how gender fits into this picture. In the next chapter we will turn our attention to women and how they experience the issues we have described in the current chapter and especially how they insist that we keep gender in mind.

5

She Works Hard for a Living

Liking it or not liking it, I have to work. If I don't work someone will die.

—(Laura)

A glimpse into Cape Verde shows a portrait of women at work: they are the buyers and sellers at the market, the caretakers of children, the fish cleaners, the water collectors, and the shopkeepers. Men are there too, but they are more likely to be on the sidelines talking with friends, playing cards, or listening to music. The women are the lifeblood of the country. Comprehending their experience fills out our understanding of Cape Verde in general (Solomon, 1992). It also allows us to see the lives of women who are often overlooked and whose experience helps us to gain a fuller picture of women in the world.

Our review in the previous chapters of the current political and economic terrain in Cape Verde has shown us that economic success or even survival is a challenge. For women this is even more true. Women in Cape Verde face many economic difficulties including high unemployment and low pay. In addition, they often are single heads of households because their boyfriends or husbands have migrated abroad. Women too are increasingly migrating abroad to search for work but the majority of Cape Verdean women today stay and do not emigrate and therefore must find a way to make a living and a life in Cape Verde. This chapter reviews the experience of Cape Verdean women with the economic difficulties that have been brought to them by globalization as well as the ways they have responded to these challenges.

Migration

Cape Verde has a long history of migration due to economic difficulties. Like much of the world, especially Africa, Cape Verde history is one of forced migration, first through the slave trade and then through the collapse of local economies, the degradation and low wages paid to people of the South, and the demand for cheap labor in the North (Potts, 1990).

Today, the diaspora (the numbers of Cape Verdeans living outside the country) greatly outnumbers those living inside the country. In 2006, approximately 700,000 Cape Verdeans lived in other nations while 421,000 remained on the islands. These numbers are only approximations, however, because Cape Verdean diaspora population estimates are difficult to obtain for three reasons. First, there are rising numbers of third or fourth generation Cape Verdeans with mixed heritage living abroad making it complicated to categorize them. Second, there are large numbers of undocumented Cape Verdeans living in other nations. Third, approximations of populations by community leaders in immigrant neighborhoods outside of Cape Verde may be overstated (Carling, 2002b).

Regardless of exact diaspora numbers, scholars agree that migration has undoubtedly decreased in recent years, while population growth has increased in Cape Verde (Carling, 2002a). One important reason migration has slowed is because nations of the Global North are increasingly altering their laws in ways that prevent immigrants from entering their labor force. Since the 1970s, immigration policies in North America and Europe have become tighter thus decreasing the numbers who migrate (Carling, 2004a). These strict policies create difficulties for young people who wish to migrate in order to increase their social mobility, as many of their parents did in the past (Carling, 2001).

Men Leaving and Women Staying

Those who leave the islands and those who stay are not just random individuals. Gender is an important factor in staying and leaving. Historically men have been most likely to leave although women are increasingly joining their ranks. Worldwide, half of all migrants (95 million) are women today. Like women all over the world, at the close of the twentieth century, more women from Cape Verde went to join husbands as more families emigrated, as opposed to single people (men) emigrating alone. In addition, single women joined the ranks of single men seeking employment abroad.

In a new country, women become the main breadwinners and send money back home to their families (United Nations Population Fund, 2006). In many countries, the "feminization of migration" has created a pattern where women go abroad, usually to countries of the North, to care for children or the elderly, to become domestic maids or migrate for marriage. For example, large numbers of women from Cape Verde now go to Italy or France to work as maids.

Other women may emigrate as they enter the global sex trade (United Nations, 2006a) and some women have found economic success moving back and forth between Cape Verde and other nations as international petty traders. Women travel to West Africa, Portugal, or Brazil where they buy goods and then return to sell the goods in Cape Verde in markets or in their own shops. Women who have family ties abroad which bring with them important networks and family support at home have found great success in these trading arrangements (Marques et al., 2001). Migration often improves women's economic status and opportunities, self-confidence, freedom, and independence although this picture is not true for all women who emigrate. Much has been written about the terrible exploitation, sexual abuse, and violence women migrants suffer at the hands of their bosses, or deceitful traffickers who misrepresent themselves as employment agents and then force women into prostitution (McGill, 2003; Ehrenreich & Hochschild, 2002).

The gender balance in emigration has changed so much very recently that in some places Cape Verdean women now outnumber Cape Verdean men emigrants (Carling, 2004b; 2002a). This change to more out-migration by women, however, is relatively recent and has occurred at the same time that out-migration, in general, from Cape Verde has declined. Overall the proportion of men emigrants from Cape Verde is still larger than that of women resulting in a continued gender imbalance favoring women within the nation. In 2006 there were over 13,000 more women than men in a population of 421,000, which included 150,000 children under fourteen, 124,000 men and over 137,000 women (CIA World Fact Book, 2006).

Remittances

Cape Verde is highly dependent on money earned by the 700,000 Cape Verdeans who live abroad in the United States, the Netherlands, Portugal, Italy, France, and Angola. In 2002, these remittances provided an estimated 23.2 percent of the Gross Domestic Product. This amounts to $321 per person per year and makes Cape Verde fifth in the world in regard to the importance of remittances to their economy (*Africa Research Bulletin*, 2006). There is some debate about how reliable these numbers are since remittances are part of a hidden economy and are difficult to measure accurately (DeParle, 2007). Nevertheless, remittances undoubtedly help at least some women who are left behind as heads of households in Cape Verde although this is certainly not true for all of the women who remain behind. Men migrate in order to improve their economic opportunities, and often continue to share their wages earned abroad with their families in Cape Verde. However, women in Cape Verde do not benefit from remittances when their husbands emigrate if the men do not return home or do not send remittances home.

This situation is common around the world. Remittances from emigrants are a huge part of many economies. The World Bank (2006) estimates that about $230 billion are sent by emigrants to their home country every year. But in many families, wives and children are left behind and money is not sent home. Many scholars argue that men's migration usually leaves women and children worse off economically, since pay is low and money sent home is minimal or nonexistent. Chant (2001) found in Costa Rica, for example, that often men who have emigrated eventually stop sending remittances altogether. This may be due to the fact that men who find work abroad are often not well paid. They may not be making enough money to send any home or they may only be able to send a small amount and feel ashamed so they don't send any. But perhaps most likely the men have established relationships with other women and are using the economic resources to support their new families.

Men's migration could improve the economic situation for the families they leave because it lessens expenses in the home. When the man leaves, there is one less mouth to feed but men are also frequently important breadwinners, especially men who have skills enough to emigrate to another nation to find work. The reduction in income, therefore, is greater than the reduction in expense. Women left alone must increase food production to replace crops men previously raised or they must increase their wage earnings to replace those previously brought home by men. Women who remain in the country must usually take sole or nearly sole responsibility for supporting themselves and their children, no small effort given the difficult economic environment.

Women who are left behind also face social pressures. Sometimes migration is used, or at least perceived, as a polite term for desertion (Elson, 1992; Steady, 1981; Chant, 2001). The woman who is left behind by an emigrating partner may be seen as a failure because she has lost her husband. In other cases, some community members may keep her under watch and report her behavior back to her husband or his family (Obbo, 1980).

Family Structure and Child Raising

Women staying and men leaving serves to intensify gender differences as well as tensions in gender relationships. These tensions are revealed in issues in marriages and child-raising. Monogamous marriage in Cape Verde is relatively unusual. The long absences and unreliable returns of the men who leave the islands make marriage and two-parent nuclear families nearly nonexistent. Instead, men who remain behind are outnumbered by women, and often maintain sexual relationships with several women at a time and have children with two or three women at a time (Chelala, 2000).

Although monogamous marriage is relatively unusual, childbearing is highly valued, whether the fathers are involved with one or with several women (Davidson, 1989). This means that many women become single mothers. Estimates are that as many as 41 percent of households are headed by single mothers in Cape Verde (Baker, 2006). In the rural areas women-headed households are even more common where almost two-thirds (62%) of households are headed by single women (CIA World Fact Book, 2008).

A woman's self-respect is tied to having children and Cape Verde has many children: 36 percent of the population is under 14 (CIA World Fact Book, 2008). About 80 percent of children are born to unmarried women. The relationships among men, women, and children regarding the diversity in marital and childbearing relationships have even generated a language to describe it. *Pai de filho* refers to "father of child" and women refer to their various partners as "my first *pai de filho*" meaning the father of my first child and "my second *pai de filho*" meaning the father of my second child, and so forth. Similarly the term *mai de filho* means mother of child and men refer to women as "my first *mai de filho*" and so forth (UNICEF, 1996).

Child raising is nearly always the responsibility of mothers alone. A large proportion of households consist of women and their children who are quite independent of men although they rely to some extent on money from the men who either live abroad or in the islands (Irwin & Wilson, 1998).Fathers, however, are not perceived

Figure 5.1 Children in the rural areas. This picture was taken at Ribeira Principal, Santiago Island.

as financially responsible for all of their children. A man's economic responsibility is usually only associated with the children of his legal wife (Finan & Henderson, 1988).

In regard to childcare beyond economic support, most men who remain in Cape Verde also do not often take responsibility for childcare and maintain that their job is to "oversee" women as they take care of children, because such work is not for men. Those men who wish to take the unusual step of participating in the day-to-day care of children are likely to suffer ridicule from neighbors and friends who see them doing women's work. Men's care of children, therefore, if it is to take place at all occurs in private. In addition, men are more likely to care for their sons rather than their daughters because they fear they might injure their daughters. They argue that caring for a boy need not be as careful or gentle as is required for daughters because girls are more delicate (*Encontro National sobre a Problematica da Mulher Cabo-Verdeana*, 1986; UNICEF, 1996; United Nations, 2006b).

We looked at what Cape Verdean women had to say about their lives and in particular their relationships with men and their children as they

Figure 5.2 Children learn to carry things on their head in Praia, Santiago Island.

work to provide for their families in a nation with severe economic challenges and an unusually low proportion of men in the population. The women we interviewed worked at a broad range of occupations. Street vendor and domestic servant were the most common categories. Other women worked as school cook, laundress, store clerk, librarian, secretary, teacher, or nurse. The women ranged in age from 18 to 65, with an average of 30. The average number of children for the mothers was three. Two women had finished university and were high school teachers. The years of education for the remaining interviewees ranged from 1 to 12 with an average of 6.7.

Stories of Economic Survival

The stories the women told about their lives reveal a dual struggle to both survive and to establish themselves as independent individuals with needs and aspirations beyond their responsibilities for children and the men in their lives. Sometimes these two goals worked together. Sometimes they experienced serious tensions between their need to survive in a difficult economic and social context and their desire to express themselves as individuals.

Descriptions of the struggle to stay alive stood out in the interviews. They often used the term survival to describe their work. Nora, a street vendor, for example said,

> In my opinion my work serves as a survival means, because there isn't any other means. (Nora)

Women spoke of the problem of making ends meet and their unstable financial situation. Diana described the difficulties of her work as a street vendor. Her job barely freed her three children and herself from hunger. She explained,

> I make a living by buying and selling. It is not sufficient but I try to resell, to do my best. Sometimes I do not sell anything and sometimes I pass through difficulties but I always try to get by; my life is buy and sell. Sometimes I do not sell, but sometimes I sell a lot, which compensates for the days I did not sell. This job does not help feed my three children [sufficiently] it just frees them from daily hunger. I have to comfort myself with sometimes selling and other times not selling. I get by only day to day (Diana).

The difficulties the women face sometimes come from political and economic sources like unemployment, the high cost of living, and the lack of dependable and affordable childcare. Nora explained how difficult it is to find work. She said,

> I would like to do some type of other work but jobs are rare so I have to be content with what I have. (Nora)

Others mentioned another problem less familiar to people in the global North but increasingly recognized as a serious and gendered social issue, drought. Francisca raises animals for a living and she said,

> I raise animals and food but it depends on the rain. Well, it is enough to survive but we need to be [more] comfortable with it. (Francisca)

The women interviewed described a situation in which they must provide for themselves and their children with little if any assistance from the men in their lives and the economic difficulties and low wages were enormous burdens to bear. They see themselves as living on the edge, surviving (or not) depending on factors far beyond their control such as job opportunities and favorable weather.

Striving for Independence

While surviving economically was important, the women also spoke of their desire to become or remain independent. Like women all over the world, the women interviewed saw paid labor as not only a means to survive but also an avenue for being independent and contributing to the community (Chen, 1995). Celine, a nurse, articulated this belief when she described the reasons she worked outside the home:

> I am one who defends the idea that women should work, in order not to depend on someone else, because the ones who depend on others will face more difficulties and often they are humiliated. My feeling toward work is that having a job is a good thing that may help us to have and lead a better life, and also it will allow us to be seen as an important contributor to society. (Celine)

This distinction between those who emphasized work as a means of survival and those who emphasized employment as an act of independence and self-expression might be seen as class based, especially since Celine is a nurse. We might interpret this to mean that working class women must focus on practical issues of survival, while more middle class women may have room to think about their work as a profession and something that brings them personal satisfaction beyond their material needs. Lota, however, who is not a professional and works as a street vendor shared Celine's view. She said,

> I think all women should work, in order to feel independent and integrated in the society. Me and many other women in Cape Verde have contributed to the development of this society, in many ways, but we have contributed. (Lota)

Celine and Lota expressed their desire to be independent in a general way and their pleasure in contributing to their community, but some women described their need to be independent specifically from men and their wish to take care of themselves. Eunice, a secretary, for example said,

> My salary is sufficient for my living standards. I have some problems as I am the one providing for my home and I have my two brothers living with me. I also have to pay 14,000 escudos [about $120] per month for

> my studies. Thanks to my work I can be financially independent and I do not have to depend on men. (Eunice)

This independence, however, is not always just a result of the desire or intentions of the women. Many women feel they must be independent—they have been forced to fend for themselves—because men are not reliable. Katia said that she was unhappy with men and wished to be independent of a man. She explained,

> I have very little confidence in men. It depends on the characteristics I attribute to men, because men do not comply with just one girlfriend. (Katia)

This reflects the tension created by the unbalanced personal relationships between men and women. The men who do not maintain monogamous relationships, or leave the island cause women to identify lines of separation between themselves and men. Helena, a woman with five children and an estranged husband who is still in Cape Verde expressed her disappointment:

> When we got married, everything was okay for some time, but then he started to have a lot of women and I got very disappointed in him. (Helena)

Both Anja and Laura's partners emigrated forcing them to become independent. Anja said,

> He emigrated and I never heard anything about him. (Anja)

Similarly, Laura stated,

> I am responsible for their [her children's] feeding because my husband got lost and never returned home. (Laura)

Anja and Laura do not appear to have chosen to be independent although they find themselves free from dependence on their partners. Other women declared they wished to be independent and took steps to remain so. Joann explained her desire to remain independent from her husband even though he owed her money for child support. She said,

> Of course I would accept nothing from him, though the court is entitled to take it and give it to me. (Joann)

Lota also chose to be independent of her husband in their case, because of his violence. She explained why she chose to separate from him and the strategy she used to get away:

> When I got married, I was too young—19 years old. The first year of marriage we lived in the seventh heaven of delight, but as it is said, there is no rose without a thorn. After having my first son, things started to get more difficult—our relationship got worse and worse, because my partner changed his behavior and he became violent and unsupportable. The situation lasted for ten years. His friends would come over almost every day with him after work. They all used to drink a lot, and go with many women and always come home late and drunk without money. Then I started to take money from him to save it. His friends didn't like the idea and they said to him that he was not a man, because he was under a woman's orders, and a lot of things like that led him to change his behavior. After about eight years under the same circumstances, the situation became so intolerable and he began to physically threaten me. That was the highest point, which led to a divorce. (Lota)

Some women said they gave the men in their lives an ultimatum. They specifically told the men to help them financially, or else leave the relationship. In these cases, they did not directly choose independence from the men, but they made a choice to choose independence, if the men did not change their behavior. Anja said,

> I do not stay with men who do not help me. (Anja)

The women interviewed appear to be independent for a range of reasons; some of them their own choice and others because of circumstances beyond their control. But regardless of the cause, independence extracts a price. Even those women who chose to be free from their partners talked about emotional costs and financial pressure. Lota described her problems after separating from her abusive husband:

> After the divorce, I lived alone for a long time, but it was very difficult to raise and educate my children by myself. For this reason, I met a new partner, [for] which I thank God because that was the best thing that happened in my life. (Lota)

Lota chose to find another man, abandoning her desire for independence, presumably in order to fulfill her emotional needs. Other women addressed their emotional needs by remaining independent from men but seeking affection and love from their children. Rita, a single mother of four children described how important her children are to her in this regard:

> They [my children] support me in everything. They aren't a burden. Without the kids things would be worse. Children don't leave us alone, making us feel sad and helpless. (Rita)

Finding a Fit between Survival and Independence

Most women in Cape Verde find themselves in a social context in which they are often unable to create satisfactory relationships with men. Men are rare because the economic situation demands that they leave the country in order to find work and then they are usually unable to maintain emotional bonds or even financial connections because of the great distances and the changes in their lives. Many women are forced to be independent, raising children on their own and working to earn money to keep themselves financially solvent.

Both financial and emotional survival is difficult and women frequently go without basic necessities such as food as well as emotional necessities like love and affection. At the same time, the women are proud of their independence and their ability to keep their children fed and to contribute to their community. Sometimes survival and independence work together as they find jobs to work for money to provide for their families and feel proud of their independence, their self-sufficiency, and their ability to take care of their children and contribute to their communities. At other times, however, the struggle for survival means the women are tied to the needs of others in ways that prevent either their education or their participation in the community. In the worse cases, despite heroic efforts they are unable to provide sufficiently for the most basic requirements for themselves and their dependents.

Finding Fault

Who is to blame? The women in the interviews explain their difficulties as largely a result of men's behavior, especially their infidelity. Some scholars have suggested that we conceptualize infidelity as a form of battering. Given the emotional pain it creates and the risk of physical damage from sexually transmitted diseases, particularly HIV, infidelity is perhaps a more important aspect of violence against women than has often been considered (Barata et al., 2005; Boonzaier, 2005).

Floretta Boonzaier conducted research on violence against women in intimate relationships in South Africa. She found that the women spoke of infidelity alongside other types of abuse more often recognized by scholars. Boonzaier says that scholars have not recognized infidelity as a form of abuse, but her interviews suggest that women understand it as abuse. Furthermore she maintains that infidelity is an indication and an expression of men's dominance over women. She writes:

By drawing upon women's own understandings of their experiences, we are able to highlight an aspect that has been previously marginalized. As theorists, researchers and activists, we might be more sensitive to "victims" and give voice to women whose experiences (of a partner's infidelity) may not have deemed them "legitimate victims." Marital infidelity is therefore constructed as a unique manifestation of abuse, but it is also a symptom of male power in heterosexual relationships.

In our interviews, women's description of their personal interactions with men is filled with anguish over their problems in their relationships with men. Some of them seemed to resign themselves to sharing men although they believed strongly in monogamy. Laura described her experience:

> I would not feel good [if my husband was not monogamous], but good or not good, you can do nothing. My husband had a lover who lived near my house. She was willing to even throw stones into my house, but once I stopped caring about her, she stopped insulting me and left me alone. (Laura)

Similarly, Paula said,

> It is his problem. Of course I would feel sad but what can I do? (Paula)

Others were less distressed but they too resigned themselves to "sharing" men. When asked how she would feel if her partner were not monogamous, Josiane stated,

> I would feel normal, because I think that is a normal thing to happen. (Josiane)

And Mindy explained her understanding of the uneven gender ratio,

> Of course I don't feel okay, but we have to share, because it does not work one man with one woman, does it? (Mindy)

Elizabeth Allen and her colleagues (2005) assert that although EMI (extramarital infidelity) causes a great deal of distress for many women, it is often normalized as an inevitable aspect of monogamous relationships. These women appear to reflect this view of their partners' infidelity. It causes them great pain but it also to some extent "accepted" as something that cannot be changed.

Some of the women, however, did not accept the non-monogamous behavior of their partners. Jocelina, for example, said that she would confront her husband. She stated,

> [If my husband were not monogamous] I would speak to him and say to him, choose between me and her, which one of us you will stay with. (Jocelina)

Even more assertively, Molly said that she would not give her partner a choice in the matter. She said,

> Since I was young, my slogan was if my partner begins a relationship with another woman, I'd leave him, I won't stay with him. (Molly)

Although the women have widely different ways of handling their relationships with men, there are two similarities in their responses. First, they often describe their associations with men as difficult because they are not monogamous. Second, nearly all of the women do not seem to include the uneven gender ratio as being the source of the problem. Only Mindy mentioned the uneven gender ratio and even she seemed to imply that it is a given, something to be accepted. Rather, they see the source of the problem as the men themselves.

Two Power Structures

Women in Cape Verde are faced with the problems of all people in Southern economies in the global system. Their situation is further intensified by the removal of men from their households leaving them to care for their children as well as themselves in an economically difficult if not desperate situation (Finan & Henderson, 1988).

Our interviews reflect the literature that says huge economic forces alter family life and gender relations in the Global South (Rai, 2002; Ault & Sandberg, 2000; Acosta-Belén & Bose, 1995). First, privatization and government cutbacks lead to a loss of jobs (Shah, 2001). Migration is one result and men have been more often the ones who go, leaving the women behind (Ludtke, 1989; Solomon, 1992). In most cases, women become head of households as well as primary income providers. Women describe how husbands emigrated or "got lost," and did not return home. Women are left with raising children, continuing household duties, and finding some type of income in order to survive.

Second, women must enter this depressed economy and find that jobs are scarce and wages are low. Ghyasuddin Ahmed (1994) writes that the labor migration and the serious gender ratio imbalance in many nations in Africa means that once poor women become mothers they become "economic slaves" because they have no choice but to work and the work they find pays so little that they are unable to do much more than survive. Many women are barely able to manage in Cape Verde. They find insufficient income through petty trading, selling prepared food, providing domestic or sexual services, or finding manufacturing jobs. The manufacturing jobs that do exist are difficult because as the demand for manufacturing of exports increases, the number of factories with poor working conditions increases (McGowan, 2007). In sum, women bear the blow of the economic arrangements of globalization (Ault & Sandberg, 2000; Rai, 2002). Survival becomes the operant word as women must search for means to endure economically and emotionally. They often used the term survival to describe their work, or they mentioned that they barely "got by."

From a social structural point of view which focuses on the global and national economic situation, the most salient factor in the women's lives is the difficult economic context that creates job shortages and low wages, which subsequently cause massive out-migration especially of men. This leaves many women faced with serious economic problems coupled with difficult personal relationships with men because of the uneven gender ratio.

The women interviewed were very aware of these problems but their working analysis focused primarily on men as the problem especially men's infidelity. They created a range of ways to cope with the problems. They worked hard to try to overcome the economic problems and they either accepted the infidelity of the men in their lives or they confronted the men. They do not, however, express an understanding of the other fundamental causes in the social structure of the economy that offers them only low wages and forces the men to leave (Albrecht, 2001). If we take Boonzaier's advice and begin to think about infidelity as a form of abuse, we are better able to see why the issue of personal relationships with men seems to be central in the women's experience, while the political and economic context remain less visible to them. In addition, if we extend this argument, as Boonzaier does, and acknowledge that when a man is in a relationship with a woman who has the expectation that they are monogamous but he is engaged in sex with other women, his infidelity is an indication of unequal power relationships. His behavior and their relationship is not only a private matter, it is in fact an aspect of a large system of political inequality based on gender. His infidelity both expresses and reproduces gender relationships where men dominate over women.

Our findings tell us that women's lives are indeed affected by the economic forces of globalization. But they also tell us that another power structure also shapes the women's lives: gender politics. From the point of view of the women, as revealed by these interviews, the global economic system takes a back seat to what they perceive to be the more salient social structural factor of gender inequality, specifically women's oppression and social disempowerment.

The organization of gender separates men's and women's experience. Men have been and still are expected to go to the ends of the earth to find a job. Women are increasingly likely to find that this expectation is also part of the social construction of femininity. But the gendered character of migration, especially in the past, has created a division between women and men in Cape Verde that literally divides them by thousands of miles.

Gender expectations and relationships also make women vulnerable to men's infidelity. Gender expectations for men require them to have intimate relationships with many women and to produce children from those relationships. Men are also expected not to worry too much about being married or being monogamous if they are married. Women, on the other hand are expected to bear the infidelity or to take responsibility for it by confronting the men and attempting to force them to be monogamous.

Gender expectations also promote the idea that women must have children and those expectations furthermore make women primarily responsible for caring for children. The picture we have of women in Cape Verde is that of never-ending responsibility and few sources of support to aid in the work to be done.

There is no doubt that most women in Cape Verde face severe economic problems as well as difficulties with the men in their personal lives. Two major systems of inequality—the global system of political and economic inequality and the system of gender inequality—shape their lives and cause them all kinds of problems. Given their difficulties, we might expect that the women would rise up in protest against their government and the world's economic institutions as well as the organization of gender and way it oppresses and exploits women. We might expect they would be calling for policies that provide good wages for men to enable them to stay in Cape Verde as well as for themselves and that they would be organizing for better education, more childcare and stronger rules regarding child support, as well as gender equality in all social arenas from public to the most private. But such a revolt has not yet occurred.

Does that mean Cape Verdean women accept their lot and choose not to resist or challenge the constraints they face? We investigated the ways that women confront the ties that bind them and found that women do indeed resist and fight back sometimes in conventional ways of protesting and organizing to create policy change. But they also do this in perhaps unexpected ways when their resistance takes the form of everyday acts of rebellion through language and dance. The next chapter (chapter 6) looks at the more conventional forms of political challenge. Chapters 7 and 8 look at the more unconventional forms of *batuku* dance and Creole as tools of expression and resistance.

Reflections on Chapter 5 Research Methods

This chapter reports the primary empirical data we collected through our interviews. We describe the social and economic difficulties women face on a daily basis and the pride women take in working, surviving economically, and beating the odds against them in a global economy. We also relate the stories of the difficulties women perceive in their relationships with men and the ways they cope with or confront the men who cause them pain. The data reported in the chapter come from interviews which were conducted with 36 local women in the community who were interviewed by student research assistants.

This chapter ties together the information provided in previous chapters about globalization and its affect on the nation of Cape Verde and the day to day experiences of the women within that context. Our findings tell us that women's lives are indeed affected by the economic forces of globalization. But they also tell us that another stratification system shapes the women's lives as well. That power structure is the one that organizes gender.

The juxtaposition of the picture of the economic context depicted in the previous chapters and the perception of the women about where the source of their problems lies, shows us a striking contrast. The previous chapters would predict that we might find women articulating a critique of globalization and a government inadequate to the task of helping women (and men) to be successful workers and parents. Instead, our qualitative ethnographic data discloses their perception of the issues. Their perception focuses on the system of gender inequality as a source of the difficulties they must address.

At the end of chapter 2 we provided a list of factors that are essential to ethnographic research. One of these was: "Make an emic perspective central, that is, priority is given to the view of the people involved—their perceptions, meanings and interpretation." The concepts of emic and etic were developed by Kenneth Pike in the 1950s. Pike (1971) wrote that "the etic viewpoint studies behavior from outside of a particular system and as an essential approach to an alien system. The emic viewpoint results from studying behavior as from inside the system" (p. 37). There has been some debate about how best to conceptualize these two types and about how to distinguish between them, but ethnographers still think about their studies as etic or emic and our goal to develop an understanding that comes from within the community of people who are participating in our research. In our study of Cape Verde, this emic perspective is revealed through the words of the women in the interviews. Although we determined the questions to be asked, an analysis of their responses allows (forces) us to look at the situation from their point of view.

In this chapter we have used both quotations and narrative to report these findings. Our voice and the voices of the respondents are mixed together to "tell the story." We began by summarizing the quotations with some narrative which guides the reader to seeing what we believed to be most important about the quotes. We then reflect on how the information from the quotes fits together to describe two power structures that shape and batter the women's lives.

In the chapter on methods, we mentioned that in interpreting qualitative data, we need to be aware not only of what is said, but what goes unsaid. This chapter is organized around those two poles. The first section describes what was said and the second section describes what was unsaid. The final section discusses what this contrast might mean.

6

Confronting Violence

The World Health Organization (WHO) estimates that between 10 and 69 percent of women from different countries have been abused by an intimate partner (Heise & Garcia-Moreno, 2002). Violence against women is prevalent in Cape Verde just as it is all over the world. Cape Verdean women are subjected to incest, rape, and battering. In addition, the women we interviewed told us that Cape Verdean women are also frequently engaged in relationships with men who are not monogamous. Some scholars have suggested that infidelity is a form of abuse (Boonzaier, 2005).

In this chapter we pursue our exploration of the problems women face in Cape Verde by investigating the widespread problem of violence against women. We were especially interested in women's response to violence against women as a political issue. International, regional, and local organizations have created forums, policies, agencies, and other structures for informing people about the issues, stopping the violence, and contending with the perpetrators and the victims of violence against women.

The United Nations has taken a lead in these programs with educational programs and international agreements obligating nations to develop and implement policies that stop violence against women. One of the most successful and well known of these is the Convention on the Elimination of all Forms of Discrimination Against Women (CEDAW) which was adopted in 1979 by the UN General Assembly. Because CEDAW defines what constitutes discrimination against women and sets up an agenda for national action to end such discrimination it is sometimes called the international bill of rights for women. The Convention defines discrimination against women as "...any distinction, exclusion or restriction made on the basis of sex which has the effect or purpose of impairing or nullifying the recognition, enjoyment or exercise by women, irrespective of their marital status, on a basis of equality of men and women, of human rights and fundamental freedoms in the political, economic, social, cultural, civil or any other field" (United Nations, 2007).

In 1992, CEDAW declared that even though this statement did not explicitly identify violence against women as a form of gender discrimination, it was implied and therefore part of the duties of nations who had agreed to CEDAW included developing programs to stop violence against women and creating reports describing the problems remaining and the efforts being made to address them (United Nations, 2007).

So far, nearly all nations in the world have signed on to the convention. Only a handful has not ratified it: the United States, Iran, Somalia, Sudan, Nauru, Palau, Qatar, and Tonga. When a nation ratifies CEDAW, it means they are legally bound to "incorporate the principle of equality of men and women in their legal system, abolish all discriminatory laws and adopt appropriate ones prohibiting discrimination against women; establish tribunals and other public institutions to ensure the effective protection of women against discrimination; and ensure elimination of all acts of discrimination against women by persons, organizations or enterprises" (United Nations, 2007). In addition, convention signers must submit national reports outlining the measures they have taken to comply with obligations under the agreement.

Regional agreements to end gender inequality have also been developed including the protocol to the African Charter on Human and People's Rights on the Rights of Women in Africa which explicitly defines violence against women as a key factor to be tackled in the struggle for women's rights. The charter asserts, "States Parties shall combat all forms of discrimination against women through appropriate legislative, institutional and other measures" (ACHPR, 2008). The charter further explains what this means:

> "Discrimination against women" means any distinction, exclusion or restriction or any differential treatment based on sex and whose objectives or effects compromise or destroy the recognition, enjoyment or the exercise by women, regardless of their marital status, of human rights and fundamental freedoms in all spheres of life. "Harmful Practices" means all behaviour, attitudes and/or practices which negatively affect the fundamental rights of women and girls, such as their right to life,

> health, dignity, education and physical integrity. "Violence against women" means all acts perpetrated against women which cause or could cause them physical, sexual, psychological, and economic harm, including the threat to take such acts; or to undertake the imposition of arbitrary restrictions on or deprivation of fundamental freedoms in private or public life in peace time and during situations of armed conflicts or of war. (ACHPR, 2008)

Violence in Cape Verde

Cape Verde is a member of the African Union and has also ratified the United Nations CEDAW agreement. However, although violence against women is a widespread problem in Cape Verde thus far the Cape Verdean government has not responded in a sufficiently effective manner.

Violence against women includes rape, marital rape, and battering. Another critical problem is child sex abuse, along with child prostitution, with girls as young as 13 joining the profession. Child sexual abuse does not exclusively affect girls. Boys are also raped and assaulted, but the majority of victims of child sexual abuse are girls. These are all aspects of the incidence of violence against women in Cape Verde.

In the past, the Cape Verdean government has ignored these problems. Today some attempts are being made to remedy the situation through government platforms and plans, and media efforts to create awareness. Violence against women and girls, such as sexual abuse, wife abuse, and rape, are now treated by the legal system in Cape Verde as serious issues. Rape and marital rape are both against the law. If men are found guilty of rape, a sentence from eight to sixteen years may be given. If men are found guilty of violence or abuse, resulting in "serious harm to physical or mental health," they are sentenced to two to eight years' imprisonment. The laws further state that if the abuse results in permanent disability or death, a sentence of four to ten years' may be given (U.S. Department of State, 2007; United Nations, 2006b). The government, however, does not consistently enforce the law.

The National Assembly amended the Penal Code in 1998, broadening the definition of sexual abuse and toughening punishments against offenders. The law protects some rights of the victims, but it does not guarantee the right of compensation. After a country-wide advocacy campaign on reporting sexual violence, there are indications that more women are willing to report violence; however this is still not necessarily the norm (Baker, 2006).

Many acts of violence against women in Cape Verde may never get to the authorities for legal consideration. The Cape Verdean government and NGOs in the nation advise women to report violence. Many women, especially women who are illiterate, however, may not know their rights or may not have the means to seek redress (U.S. Department of State, 2007; Commission on Human Rights, 2003). In addition, 70 percent of women in Cape Verde live in the rural areas far away from shelters, hospitals, protection, and assistance. Not knowing where or who to turn to, in addition to difficulties with transport and money pose problems for women in the rural areas who want to report violence or might want to report violence to the authorities if they knew that was an option.

Gender ideologies also prevent women from reporting violence. Keeping the family together is seen as a woman's responsibility and no amount of suffering should cause the woman to report violence and ask for court intervention (Adande & Kuye, 1986). Women are told no good will come to the child, to the husband, or to the wife herself if the court intervenes (U.S. State Department, 1989–1997). In some communities, women know that turning a family member in to the police is decidedly not in their best interest. They may face the additional threat of physical and psychological retaliation against them if they press a rape or abuse charge (Halim, 1994).

Although services exist to support women who have experienced violence, such as legal counseling, psychological care, and family courts, in practice these neither ensure punishment of the perpetrator nor effectively prevent future violence. Women explain that police overlook the legal complaints they file against their partners. When women do file complaints, the police and the courts often delay in responding to the complaints (U.S. Department of State, 2007; Commission on Human Rights, 2003).

25 Year Delay

Under Cape Verde law, women are guaranteed equal rights with men, including rights under family law, property law, and in the judicial system. Government officials claim they are committed to improving women's lives, economic opportunities, and access to education. For example, Cape Verde was one of the first countries to endorse CEDAW, which was signed by Cape Verdean leaders in 1980. The convention became official in 1981 and Cape Verde was supposed to submit its first report within a year, documenting the progress made on eliminating discrimination and improving women's lives (United Nations, 2006b). However, it wasn't until almost 25 years later that the first report was finally submitted. Government officials explained the late report was due to bureaucratic challenges, lack of staff, and lack of budget.

The government organization also established an agency to address women's issues, the Institute on the Status of Women in 1994. Its opening, however, was also delayed due to lack of staff (United Nations, 2006b). In spite of the 25 year delay, officials claim they are still committed to assisting women and integrating gender into development and a number of policies and plans have been adopted to achieve that goal (United Nations, 2006b).

In addition, when Maria Cristina Lima, minister for the presidency of the Council of Ministers, State Reform and National Defense, was chided by the United Nations CEDAW committee for the lateness of the report, she pointed out that Cape Verde had made great strides for women's rights. She said her country has come a long way since 1975, when a woman could not be a judge and had to seek her husband's permission to conduct trade (United Nations, 2006b).

National Plans for Improving Gender Equality and Economic Opportunities

One important way of ensuring that violence against women is adequately addressed is to make certain that women's voices are heard at the highest levels of political power. In Cape Verde, this would mean bringing more women into the national parliament. In 1996, The National Action Plan for the Advancement of Women established gender mainstreaming as a feature of its policy planning. The United Nations defines gender mainstreaming as,

> a globally accepted strategy for promoting gender equality. Mainstreaming is not an end in itself but a strategy, an approach, a means to achieve the goal of gender equality. Mainstreaming involves ensuring that gender perspectives and attention to the goal of gender equality are central to all activities—policy development, research, advocacy/ dialogue, legislation, resource allocation, and planning, implementation and monitoring of programmes and projects. (OSAGI, 2008)

In 2000, the Cabinet of the Prime Minister of Cape Verde, the Institute on the Status of Women, and the United Nations Population Fund developed a program to promote gender mainstreaming by integrating gender into development policies in the nation. The Institute on the Status of Women was later renamed the Cape Verde Institute of Gender Equality and Equity, aiming to attain gender equality and women's empowerment in various areas. The institute created a national plan from 2005 to 2009 specifically to fight violence against women, improve women's economic opportunities, and increase participation in political life. Women now hold one of the five positions on the Supreme Court and 33 percent of other judicial posts, and 32 percent of diplomatic slots (United Nations, 2006b).

Despite these important steps, however, the participation of women in elected office remains quite low in Cape Verde. Women account for less than 20 percent in political posts and less than 15 percent of members of parliaments (United Nations, 2006b).

Violence, Shelters, and Media

The government has recognized that violence is a huge problem but only minimal practical response has been forthcoming. For example, one shelter was opened to assist women in 2004 in the capital city, Praia, which has a population of almost half a million (423,613) (CIA World Fact Book, 2008). In 2005 and 2006, the government established centers in hospitals to receive victims of violence. During the first 6 months, the centers received 250 women (United Nations, 2006b).

The media have been brought in to support these efforts. The weekly radio program *Femina*, for example, has aired since 2003 and in 2004 the program *Mudjer* was aired by the Organization of Cape Verdean Women. Both programs discuss HIV/AIDS, breastfeeding, worker's rights, and reproductive health. To increase awareness about violence against women, a program was televised called" Breaking the Silence," which led to an increased number of phone calls made by women to the police (United Nations, 2006b).

Trying to understand the source of violence is one with which many nations have grappled, including Cape Verde. Is violence a problem that is solely tied to families? Does the government play a role in violence? Is violence connected to the economy? December Green (2000) explores these three questions for understanding violence against women in Cape Verde. The first section discusses the first of these: the idea that the family is the central location of violence against women.

Families and Violence

Because of women's critical reproductive and productive capabilities, they are both highly valued and jealously controlled. Cape Verdean women, for example, are idolized for their fertility at the same time they are restricted because of childbearing abilities (Nhlapo, 1991). They are revered for bearing children but mothers or potential mothers must follow strict rules such as staying in or near the house. Women who leave their house may be seen as loose women. Women who go against tradition and remain childless are viewed as threats to the most cherished values: fertility, motherhood, and families (Green, 2000). Women in Cape Verde, for example, who choose independent actions such as: living on their own; putting career before motherhood; choosing not to have a child; or attempting to remain friends with a man instead of having a sexual relationship—are not understood and may be regarded as strange and selfish.

Living within these rigid and contradictory sets of rules is difficult for women and men and can result in violence against women. When abuse happens, women are blamed for supposedly provoking the assault and they are told to accept their punishments for doing wrong. The media, the church, the schools, and families create and endorse gender ideologies that keep women in their place, enduring the abuse. Institutional barriers for stopping violence against women in the form of delayed or nonexistent help from police or government reinforces the ideology of blaming the victims all over the world (Rafter & Stanko, 1984). Cape Verde women assert that in their communities the police regularly ignore their calls for help against their partners in cases of violence (United Nations, 2006b).

Government and Violence

Violence against women often takes place within intimate social circles, especially families at the hands of men who are the husbands, fathers, uncles, and friends of the women they abuse. Some people argue that the problem must be solved at this micro level by looking at relationships within families and changing people within them. Others argue that the government must be drawn into the struggle and that the problem is broader than families and personal relationships. The legal system, for example, must be changed in order to successfully address the problem of violence against women (Green, 2000).

However, even those who believe the government could be a solution are not always sure it will be one. They note that governments frequently do not work in women's interests. The ideologies governments promote and the laws they maintain often do not stop violence against women and may in fact encourage it through both action and neglect. Gender ideologies guide the manner in which women are treated by police, lawyers, judges, and juries. Government policy in Cape Verde, for example, reflects an ideology that regards violence in the home as a family matter, not as an issue that the state should handle (U.S. State Department, 1989–1997). In addition, the legal system is not well respected by a significant minority of Cape Verdean citizens: 28 percent of the public regard police officers as involved in "some corruption," and 19 percent feel the same about judges and magistrates (Ames et al., 2003; Baker, 2006). Women have few chances against their perpetrators without responsive law enforcement, community involvement, or cultural conviction (Copelon, 1994).

Beyond the abuse itself, violence between partners and the disruption of partnerships can cause further economic problems. When a man and woman separate, women typically do not ask for and do not receive child support. Women may be hesitant to use the court system based on the belief that men should not be expected to help pay or care for children after a separation (Urdang, 1989; Maboreke, 1991). In chapter 5 for example, one of our participants, Joann, explained that she did not ask for child support from her estranged husband, though she was entitled to it. She is not unusual and the result is increased poverty among households headed by single women: 32 percent of woman-headed households are poor and 68 percent are very poor. In contrast, 26 percent of man-headed households are poor and 14 percent are very poor (Baker, 2006).

Throughout Africa, the legal process in cases of violence against women, and especially rape, often work against women. Difficulties in calling the police and making formal reports, language barriers, mistakes by medical workers and, court inadequacies all serve to hinder women's ability to use the laws to protect them from violence. When a woman calls for help, police may often fail to respond at all, or assistance may be delayed when calls from privileged areas take precedence over calls from disadvantaged neighborhoods (Human

Rights Watch, 1995). Because police officers often lack special training, sensitivity, and counseling techniques, they are not equipped to interview victims and document the rape appropriately. Language also may play in part when the police officers do not speak the dialect language of an area to which they have been assigned. Utilizing interpreters often only increases the victim's ordeal or leads to inaccuracies in the report (Taylor & Stewart, 1991). Police may be suspicious of false allegations and stereotype all rape victims as seeking some type of revenge toward their mates. Especially for rural women, just getting to the police station can be a problem due to unreliable, infrequent, and expensive transport (Green, 2000).

The medical exam is another area of insufficiency. Some physicians are not well trained to collect evidence or support the rape survivor. It is often the case that the physical examinations are not properly conducted and evidence may be badly taken or incomplete (Green, 2000).

Finally, the court system is a problem. Lawyers are not always accessible. Women, especially rural women, are not only the least educated but also the least able to afford lawyers, and the most likely to be intimidated by the impersonal, formal, and imposing court process (Maboreke, 1991). If a case gets to prosecution, in many countries, the criminal justice system is already filled to capacity that prosecutors are pushed to be selective in the cases they take. In 1998, "reforms to strengthen an overburdened judiciary were implemented. Between 2000 and 2005 there was a 15 percent increase in the number of judges and magistrates" (Baker, 2006, p. 503). But delays and uneven prosecution remain.

The view that family should be the focus of efforts to stop violence against women is flawed because larger social institutions such as the government, the media, and health care providers also need to be called upon to address such a prevalent and difficult issue. The government in Cape Verde like those of many nations, however, has failed to live up to its potential for stopping violence or assisting women to leave violent situations. One answer to why governments have been so ineffective is that they exist within an economic context that does not value women and leaves them dependent and economically needy in a way that hinders their ability to avoid violent relationships and extricate themselves from those that become violent. In addition, women are not well placed in the political system and therefore, have little leverage in creating better, more responsive systems for stopping violence against women.

Economics and Violence

The third question Green (2000) addresses in her review is how violence against women is connected to economic relations of power. To begin this assessment, Green (2000) discusses the concept of patriarchy and in particular the expression of patriarchy in the workplace. She cites the work of Myra Strober (1984) who defines patriarchy as, "a set of personal, social and economic relationships that enable men to have power over women and the services they provide." The workplace is one of the most important places where this kind of exploitation takes place.

Women all over the world earn significantly less for their labor than do men. Many scholars have noted the importance of gender segregation of jobs in explaining this gap, but Barbara Reskin and Irene Padavic (2001) contend that gender segregation of the labor market is not a sufficient explanation of why women's wages remain lower than men's. They suggest that gender segregation of jobs is a key factor, but even more important is the politics of gender—the power differential between women and men. Reskin and Padavic explain "the basic cause of the income gap is not sex segregation but men's desire to preserve their advantaged position and their ability to do so by establishing rules to distribute valued resources in their favor" (Reskin & Padavic, 2001, p. 258). Some men have power to make the rules and they wish to maintain both their power to write the rules and their ability to benefit from them. According to Reskin and Padavic, the cause of low wages for women's work is power differences and the result is less access to resources for women (Aulette et al., 2009).

Green (2000) suggests that this power differential is at the root of both the economic gap and the political gap between women and men which makes women vulnerable to violence. Even when an economy is running well, women often do not make enough money to be independent of abusive men. In addition, during periods of economic, political, and social disorder, violence increases and the "abnormal" functioning of the economy can be accompanied by increased violence against women (Green, 2000). Economic pressures on urban and rural economies in Cape Verde have led to a flight to the cities or to other countries abroad and as we noted in the previous chapter, typically men are the ones that leave but women are increasingly likely to be migrants as well. Violence may increase when women challenge men about little support from remittances (Brettell & Sargent, 1993; Elson, 1992; Moore, 1988). In addition, men may feel their role as breadwinner is lessened if women are contributors to the family budget, leading to frustration and loss of confidence. They may participate in a variety of activities to re-establish their endangered masculinity such as aggressive activities or abusing women (Brydon & Chant, 1989; Campbell, 1992). Sylvia Chant (2001) calls this a "crisis of masculinity." Men's source of identity and power are affected by changes in the economy and women joining the work force. Men experience a loss of control in the family, where they are no longer the head decision maker (Chant, 2001).

Indeed, the changing world economy has impacted work lives of women and men and it has altered gender relations between women and men. The ways people go about their daily lives, their roles, and relationships with others have all been reshaped by globalization. Betty Potash looked at these changes in family relations in Africa and notes:

> The conversion from subsistence to a cash economy, labor migration and male absenteeism, new bases of status differentiation and a general devaluation of women's subsistence activities are some of the developments that have affected women's family relationships. Urbanization, new legal codes of marriage, and new ideologies and religions are other factors associated with modifications in women's situations. (Potash, 1995, p. 68)

Although some might view globalization's tendency to push women into the paid labor market as an advantage for women, women's entry into wage labor may merely alter their subordination but not eliminate it (Green, 2000). What kinds of jobs do women find in the global economy? The establishment of multinational factories and the increase in production for external markets has often directed women into subsistence and petty goods production instead of wage labor. Cape Verdean women try to survive through selling prepared food, vegetables, and candy in addition to working as domestic maids or prostitutes.

When women do find waged work in globalized economies, their jobs are poorly paid, for example piece rate, while men more often find jobs that are paid on a daily basis. In addition, the reproductive unpaid labor of women in the home and in the fields continues to enhance the productivity of men, keep the masses of people alive through subsistence work, and most significantly increase the profits of the corporations (Bujra, 1986; Drew, 1995).

But how are these economic factors related to violence against women? Suzanne Choi and Kwok-Fai Ting (2008) have developed the concept of imbalance theory to explain physical violence against women in intimate relationships in South Africa. On the basis of their review of the empirical data of several studies, they assert that three types of imbalances in resources and power can occur that can contribute to violence against women in intimate partnerships: dependence, submission, and transgression:

> The dependence hypothesis suggests that economic dependence of the wife will lead to more violence....The submission hypothesis suggests that violence will increase due to the submission of women in male-dominated families. Finally, the transgression hypothesis argues that men in female-dominated families will use force to punish their wives for supposedly transgressing the gender norm of male dominance. (p. 834)

As African economies are increasingly marginalized, more pressure than ever is placed upon women to find survival strategies, many of which may require them to adopt roles that step outside conventional ideas about appropriate feminine behavior. We can imagine seeing an increase of blaming women who press traditional boundaries. Other women will be unable to support themselves and will be more dependent on their men partners making them less able to leave abusive situations. Or, they may be forced to submit to his authority and abuse as the only possible means to survival. In these contexts, violence is rooted in economics. Women are caught between two powerful forces. Africa's shifting economic fortunes are forcing women to enter wage employment, which could be a source of power and independence, but their jobs are not reliable and their wages are low. Women, in addition, are held responsible for the negative consequences of men's migration and the loss of African traditions. Women then experience violence from their partners or from other women. Whether it is rape or wife abuse, we can see violence against women is intrinsically bound up in and central to the running of a capitalist economy in a global system (Green, 2000).

The family, the state, and the economy all play roles in violence against women. All have failed to take sufficient action to improve women's lives, and in fact may play a role in the increased incidence of violence against women. With a lack of jobs (66% of Cape Verde women are not in the paid labor force) a lack of help from government, and being overwhelmed with household duties, women have few places to turn. As a result, Cape Verdean women and women across Africa are coming together to find solutions to these problems through public and private methods. The next section looks at ways women are responding to violence.

Taking Action against Violence through Engagement and Disengagement

Women take action themselves in a variety of ways. One way is through a public and collective approach. This is referred to as the *politics of engagement*. Women and their respective clubs or organizations attempt to change laws or change governmental policies of violence against women. A second way women take action is through a private approach. This is referred to as the *politics of disengagement*. This is the management of alternate expressions of power. In this private approach, women exert power in quiet and subtle ways to protect themselves and their families from violence (Green, 2000).

Through the *politics of engagement*, African women are attempting, with only some success to make changes through the state to address problems of violence. One strategy women use to wage this battle is to form associations and networks such as rural cooperatives, occupational associations, urban business enterprises, church and entertainment groups. These groups use a range of strategies and analysis to end violence, from self-help and social service oriented ones to those with a political focus which seek to challenge and alter social institutions such as the police, the laws, and the media (Green, 2000).

NGOs in Cape Verde are focused in particular on fighting against sexual abuse and violence through such organizations as The Association in Support of Women's Self-Promotion in Development and The Cape Verdean Women's Organization. In addition, the Women Jurists' Association offers free legal support to women suffering from discrimination, violence, and abuse. The association has attempted to seek legislation to create a special family court to handle cases of violence. As of the end of 2007, however, legislation still had not been enacted (U.S. Department of State, 2007; Commission on Human Rights, 2003).

In our research, we explored the ways women see the problem of violence and their response to violence in the form of the politics of engagement. At a women's rights conference in a rural town, we distributed 50 questionnaires (see appendix E) that asked women about their experience with violence and the efforts they were making to address those problems.

Coming Together to Talk about Violence

> *I have learned that if I act, my life will improve. I have learned positive and creative things, experiences and useful knowledge.*
>
> —*(Ana)*

Cape Verde women lawyers organized a series of women's rights conferences on the island of Santiago in 2001. A small town, where a high number of known abuse cases had occurred, was used as a venue for the conference. The conference was attended by 50 women and questionnaires were distributed at the end. The title of the conference was "Training of Trainers: A Conference on Women's Rights." The objectives of the conference were to give women a confidential opportunity to talk about taboo subjects—women's rights and violence in the home. The second objective was to have a "training of trainers." The conference was designed to train the invitees in the area of women's rights and conferencing skills. In turn, the final objective was that these women would create their own future ways of promoting women's rights, in their own communities.

The day long conference included warm-up activities and games, a presentation by the keynote speaker, questions and discussion, coffee breaks, additional speakers, lunch, and evaluation. The keynote speaker was a woman's rights lawyer. The other two speakers represented local NGOs that aimed to increase economic opportunities for women. The three women speakers discussed the problems of violence and rape and where victims can turn to for help. They also spoke about women's rights in the areas of marriage, divorce, child support, and violence. In addition, the speakers addressed women's self-confidence, decision-making, and the importance of paternity and father participation. The three speakers were chosen for the conference because of their known work and reputations for assisting women throughout Santiago Island.

This initial conference led to an island-wide and nation-wide movement of organizing women's rights conferences. One woman who attended the conference subsequently organized two additional conferences in small villages on Santiago Island. All three conferences were funded by a grant from USAID. A fourth conference was also organized which the local government funded alone.

Other women on islands of Santo Antão, Boavista, and São Vicente organized additional conferences. These women are examples of those who are pushing for change in Cape Verde, and are demanding more—more training and education, more opportunities and better jobs as well as an end to violence. Carter attended this first conference and asked women questions about what problems they faced and what they were doing to address those problems.

Their assessment of problems clustered into four categories. The first of these were prejudiced and negative attitudes about women which they believe were widespread. Discussions in the sessions as well as the comments by the women about the conference and why they were there revealed that they believed that a major feature of the culture of Cape Verde is women's second class status at work, in political institutions and in families.

The second key problem identified by conference participants was a lack of support from personal or institutional sources. They mentioned that the police did not help them nor did their partners. They believed that the many problems women faced as individuals and as a group were being faced alone, without assistance either from people close to them or from community services.

The third area was their concern that finding well-paid jobs, or any jobs at all, was difficult. Furthermore, they felt that inadequate economic resources caused them to be overly dependent on men. Related to this they wrote about a lack of educational or training opportunities for women that might help them to find better jobs.

The fourth area about which they were concerned was violence against women. Some were troubled by the problem in general or for others who were close to them. But some women had been victims themselves. One woman gave a simple and frank summary of her commitment to the conference program, "*I am one of those women who have been raped.*"

All of their responses echo what we have described in chapters 4 and 5. The women see economics as an important issue. They are concerned about jobs and they are aware of their vulnerability in the labor market that demands that they have better education and offers them few opportunities. They also identify the government as a potential source of help but one that has not been sufficiently supportive in providing education, job training, and police protection. But like the women we heard in chapter 5, they are also concerned about gender inequity and injustice in their personal relationships with men who are also not supportive and often violent. The women are concerned about both the problems in these relationships and their lack of independence from men who cause them problems.

The women also had ideas about what could be done when we asked them how to address the problems they had identified. They wrote that job training and better employment opportunities were essential. They also suggested that the government, especially the courts and the police should be more involved and provide them with assistance and protection. In addition they would like to see stronger organizations such as jurist associations and national women's organizations. And they would like to have more opportunities for finding out about these kinds of organizations and networking within them. They were especially concerned about obtaining information about women's rights and institutional resources if they needed help. The women also wanted to see more partner and family involvement in women's issues. They were seeking more responsibility among men and a greater involvement of men in their children's lives. And they also wanted their families and partners to be more supportive in their quest to gain self confidence.

Putting Knowledge into Practice

We also asked women how well the conference addressed these concerns and supported their own views of the reforms that were necessary to solve the problems of women in Cape Verde. We asked, "What did you learn from the conference that you can apply to your life?" Women explained that they had learned how to take better care of themselves and their family and they gained more information on women, men and children's rights. They also learned about what they might want to change in their lives and how best to go about making those changes.

Women explained they gained information which can help them in a range of ways. Ana shared the information she had gained mentioning that the conference had helped her in a direct practical way as she was in need of a divorce.

> I learned things I can apply to my life regarding violence and women's rights. What I learned today will be applied to my life and my work. I learned how to set in motion, a legal divorce. (Ana)

Dulce's assessment suggested that her gain had been less specific but the conference had provided her with important general information and insight. She said she realized she can focus on improving her own life, rather than allowing men to direct her life.

> Today I have learned things about my daily life, and of my children. I have learned a lot regarding success for my family and social life. I have learned that men are not God. (Dulce)

Maria's future seems to be clearer as a result of attending the conference. She said she had learned from the experiences of other women, and she had also learned that she had knowledge that she could share with other women:

> I have learned to defend my own interest as a woman of the community. I have learned to live better and to take care of my life. I learned to respect my self and my dignity. After the conference, I began to trust myself more, like a free woman who has a right to everything. I have learned a lot and now I can foresee my future more easily. I learned things that I can explain to other women. I can apply to my life the experiences of other women. (Maria)

Their observations about the conference suggest they are eager for opportunities to talk about their experiences and to gain practical knowledge that can help them cope with difficulties or change their circumstances. They also reflect a deep commitment to finding solutions as a community for themselves but also their children and other women.

Reflections on Chapter 6 Research Methods

The data in this chapter come from several sources. One set of information was secondary sources in our review of the existing literature on violence against women in Africa and Cape Verde. We have also examined primary data sources. The first of these is official documents from the United Nations, the African Union, and other sources about policies regarding violence against women in Cape Verde. The other primary data sources were a questionnaire distributed at a conference on women's rights in Cape Verde and field notes taken by Carter as a participant observer. Together, all of these sources of information provide us with a picture of how Cape Verdean women experience the problem of violence.

The documents provide a more formal description of public policy telling us what the government has or has not done. The descriptions of the answers the conference participants provided in the questionnaires at the conclusion of the meeting provide a view of the issues from the individuals who are simultaneously most directly affected by the problems and most essential to finding solutions.

As we discussed in chapter 3, conducting research cross culturally is especially challenging due to differences in language, interpretation and cultural values between the "researcher and the researched." The data collection for this chapter was particularly difficult due to language and especially language dialect. Language dialect in Cape Verde for example is a marker of status. Creole from the northern islands is said to be more sophisticated and sometimes incomprehensible to those living in the south. Creole from the south is said to be more traditional and likewise sometimes incomprehensible to people living in the north.

This aspect of language affected our data collection at the conference and illustrates one of the outside barriers we faced as researchers. For example, the key note speaker spoke in a dialect from the north to an audience who were mostly from the south and particularly from the rural areas. The women who attended the conference were most likely less well educated than the key note speaker and therefore may have been unable to grasp some of conceptualizations she gave of violence and women's rights. Finally, the questionnaire we distributed was written in Portuguese, while most of the women were fluent in Creole, and not in Portuguese.

In spite of these problems, we feel that we have uncovered some sense of women's experience of violence in Cape Verde and some of the ways women address issues of violence. We have also uncovered some of the difficulties of gathering information in research in a diverse setting.

7

Batuku Dance as Resistance

Its significance lies in the dancing gestures and the word meanings, including satires and complaints; things that otherwise wouldn't be said.

—*(Jandira)*

Cape Verde is a country affluent in oral traditions. These include proverbs, riddles, stories, songs, as well as popular beliefs, myths, superstitions, and games, which can be heard and seen when walking through the streets. One of the most popular oral traditions is the *batuku* (Creole) (*batuque* in Portuguese) song and dance (Hurley-Glowa, 2001). Men might clap their hands to the sound of the drum in the *batuku* dance or play an accompanying instrument, but it is primarily a woman's dance (Máximo, 1998).

During colonialism, King Manuel I of Portugal passed a law prohibiting the dance, along with another type of dance, *funana*, that were both said to be "too African" and "too primitive," "noisy," and "indecent" (Reis, 2006). The dance was one form of resistance to cultural imperialism throughout Portuguese rule (Máximo, 1998). Colonial powers argued that *batuku* had to be condemned and restrained because it was "rehearsal for rebellion" (Reis, 2006, p. 46). Offenders could be fined or even sent to prison. *Batuku* continued, however, underground in people's homes.

Explanations of the origins of *batuku* vary. Some argue it originated during slavery, when masters raped women. Others believe that it came from women's grief over the loss of a child, husband, or other family member while living under the harsh conditions of slavery and colonialism (Irwin & Wilson, 1998). The dance form undoubtedly came from the continent and was brought by slaves to the islands. It is the oldest documented musical tradition in Cape Verde (Fryer, 2000; Palmberg, 2002; Hurley-Glowa, 2006). Today, women chant about tragic events throughout history, such as the famine in the 1940s, which killed more than 50,000 Cape Verdeans. Or women sing about the work contracts in the plantations of São Tomé during the 1970s, which resembled a form of twentieth-century slavery. Other lighter topics include giving compliments to party-givers, offering matrimonial advice, warning against loose behavior, and expressing criticism against politicians. *Batuku* is performed at a wide array of social events including farewell parties, baptisms, local celebrations, and weddings (Máximo, 1998).

Batuku consists of a rhythmic beating and a call-and-response; the singer calls out and others respond. Words are often made up as the song progresses. The dancing is guided by makeshift drums which provide the rhythm—a rolled up *pano* or cloth held between the legs, sometimes covered with plastic and pounded with the hands. Women sit in a semicircle and sing, while one woman, wearing a shawl around the upper thighs, dances in the middle of the circle (Lobban, 1998). The woman dances slowly at first with rotating hips, and then faster until there is a peak of dancing, cloth pounding, and chanting (Irwin, 1998). Some women, in order to illustrate their skills, put a bottle of an alcoholic drink on their heads while dancing. At the end of each song the dancer selects the next dancer, by giving her the shawl, or placing it around her, and leaving the circle (Máximo, 1998).

After independence from Portugal in 1975, changes were seen regarding musical and oral traditions. Individuals attempted to revitalize and promote what they considered to be truly Cape Verdean. This was part of an effort to build a new national identity and discard the colonialist and racist oppression throughout the islands (Lobban, 1998). In the spirit of "re-Africanizing" Cape Verdeans, the new government encouraged both the study of *batuku* through research on oral traditions and the recording by musical groups. OMCV, the national women's organization, promoted *batuku* festivals and politicians often hired *batuku* groups to perform at their conventions. Today, it is broadly accepted as an authentic Cape Verdean tradition (Máximo, 1998).

Batuku songs are often made up on the spot, but they also relate to certain themes that are traditional. Those themes remember and express the difficult history of Cape Verdeans. The following excerpt from a *batuku* song reveals a number of significant themes—colonial neglect, forced labor, famine, and migration. It also reflects a hopefulness and a reference to the belief that in spite of these hardships, *batuku* can bring joy, excitement, and spirit:

> ...The soul is calling me
> To dance batuku.
> There were dozens and dozens of people
> Buried in a common grave.
> Hundreds and hundreds of people buried in a shroud of stone in the disaster
> Of the Assistencia.
> Thousands and thousands of Cape Verdeans
> Forced to labor in São Tomé,
> Some were burned in the lava of the volcano.
> The body dies but the spirit stays.
> The soul is the strength of the batuku,
> In the time of famine,
> In the sharing of excitement,
> In the longing for the son gone away,
> Batuku is our soul.
> Feel it, my children,
> Those who love us, love batuku.
> Batuku is our soul!
>
> (translated by Felisberto Vieira Lopes—1964,
> quoted in Cape Verde Web page,
> University of Massachusetts).

During colonialism, *assistencia* was the name of the Portuguese governments' soup kitchen and welfare building in the capital city. The walls of the building were made from cheap materials. The building collapsed in the 1940s, crushing hundreds of people. The incident is a true historical event that has become a metaphor for colonial neglect and the inhumanity of powerful institutions in their treatment of the people of Cape Verde (Solomon, 1992).

Personal Observation of *Batuku*

Carter participated in *batuku* dance a number of times—at weddings, parties, and just playing around with the neighborhood kids. Even a small effort put forward on her part to try the dance was warmly met with encouragement and appreciation, laughter, and gratitude by Cape Verdeans. There are many opportunities to dance *batuku*. Though *batuku* is nearly always danced during a big event such as a wedding or party, it can be danced at anytime, informally and spontaneously. Each island and town has its own seasonal festival; in addition there are celebrations on Catholic Saint's days. *Batuku* is danced informally at all of these festivals. Hurley-Glowa (2006, p. 86) writes, "You rarely meet a Cape Verdean who is not an active musician or dancer. Everyone shares in the activities." The *batuku* is danced at the same time that other activities are going on—the music playing and people eating, drinking, or going for a swim. It serves as a constant backdrop.

Usually the men sit on the outside of the circle, clapping hands and watching. Children also watch. Sometimes the shawl is given to a young girl, and she comes to the middle of the circle. After she completes her dance, the crowd claps and cheers in approval. *Batuku* is something that nearly all girls learn. Though it might be performed at weddings or baptisms, it is not connected to religion. It is seen as a celebratory dance at a happy occasion, but when hearing the words expressed, it is evident that it is also a time where women express the difficulties in their lives.

Expressing Strong Feelings

Our field notes told us that *batuku* was an important feature of the lives of women in Cape Verde. We were, therefore, interested in *batuku* and asked women to describe their experience with the dance. They stated that the words expressed were the most important part of the dance and provided vehicles for the expression of strong feelings. The words allow them to speak of important issues in ways that express deeply held beliefs and experiences. Filipa in particular uses freedom to describe the *batuku*. Women do not have many opportunities to be free or feel free-spirited. The *batuku* is one of few opportunities to let it out. The participants explain that *batuku* is a form of expression that can address a broad range of issues but its most salient feature is the strength of emotions behind that expression. They use the word strong in association with their description of the dance and they describe deeply held feelings such as freedom, love, and hatred. And they note that the most important events in human life such as birth, death, hunger, and subjugation are topics appropriate to *batuku*:

> It's a kind of rhythm in which people can really express what they feel especially in terms of freedom, euphoria, and other strong statements. (Filipa)

> Batuku came from Africa. But in Cape Verde, it took a special shape, and it was used to express general feelings—love, hatred, departure, cheating/infidelity, death, birth, arrivals, drought, hunger, subjugation, etc. (Lenira)
>
> The words are strong. People use batuku to criticize the society and/or to show their happiness in relation to specific events. The body movement is also expressive and shows people's talent. (Jorge)

Issues of Power

The specific themes in the song are important because they describe the injustices they had suffered, their opinions of the sources of the oppression, and their feelings about it. The topics cover a broad range of issues. *Batuku* reveals difficult working conditions, criticisms against politicians or society, and feelings of joys and sorrows. However all of the topics touch on questions of power, inequality, and resistance and solidarity with others in similar circumstances.

Celeste said Cape Verdeans could use *batuku* to express their difficulties of a poor life and poor work conditions:

> Because it was something inside them that they carried since they were a child and they also wanted to show their suffering regarding working conditions, life conditions, etc. (Celeste)

Eneida and Gilda argue that it not only exposes the victimization, it also exposes the perpetrators of injustice:

> It represents the way people work as slaves. It also criticizes the society—politicians, injustice in all sense, power, abuse, etc. (Eneida)
>
> According to some people, it was a way people could criticize society, economically, politically, and socially. (Gilda)

Leidiza described another significant theme of the *batuku*—the possibility of hope for change in their circumstances:

> It has another theme, which they sing when they are working in the field—the cornfields, in order to give them strength and hope; they sing together while working hard in the field. (Leidiza)

Solidarity under Colonialism

European colonial powers considered the batuku sinful, immoral and barbaric (Herbison-Evans, 1988). But this didn't stop Cape Verdeans from dancing. Participants explained that batuku was part of their national and continental (in solidarity with all of Africa) cultural identity and furthermore, it represented a form of resistance to outsiders, an expression of anti-colonialism. Participants also explained that the dance set them apart from others at the same time it expressed the unity they felt with their peers. Bela and Jano, for example stress that batuku was linked with national identity and Jano specified that batuku represented the African part of Cape Verdeans:

> It identifies Cape Verdean people. (Bela)
>
> It makes part of Cape Verdean culture. It shows the African aspect of Cape Verdean people. (Jano)

Several participants said the *batuku* symbolized, especially during the historical period of colonialism, both the suffering as well as the resistance during those times. Their descriptions note the historical continuity *batuku* represents linking their past with the present and allowing them to preserve that history. Their comments also suggest the power relationships *batuku* represents as they use political language to describe the intention of *batuku* as a protest; a revolt; a fight; a way of showing solidarity with other people under colonial rule, especially in Africa; a way of standing up to colonialism:

> It represents our past, our suffering, out fight against colonialists. It represents our culture. (Jorge)
>
> Batuku was a kind of protest to Portuguese culture imposed throughout the archipelago. (Bela)
>
> Batuku came to Cape Verde from Africa during the period of colonization. This kind of dance allowed them not only to enjoy themselves but also to demonstrate revolt towards their owners. (Jano)
>
> To challenge the colonial authorities, a way to protest. It was the same to say, "you can kill me, but you can't kill my insides, my culture." It was inside of them, one cannot kill one's soul. (Julia)
>
> It was a way to express their feelings and enjoy themselves at the same time. It was also a way to promote ideas against colonization. (Isabel)

Eneida further describes the political nature of *batuku* as it allows the dancers to retain their personal identity and personal expression under a government that prohibited freedom of speech:

> Because through batuku they could express their feelings, they could keep their culture, even in secret. (Eneida)

Passing on Culture

Batuku has been passed on as a form of resistance through many generations. When asked how they learned about the dance, some participants said they didn't remember specifically learning it during childhood but recalled seeing it during holidays or festivals. It was just part of the social surroundings:

> I learned batuku inductively. Nobody told me any story related to it. But in my neighborhood people used to dance it where there were parties like baptisms and weddings. (Eneida)

Others learned about its meaning and significance when they began to study the history and culture of Cape Verde:

> I don't remember learning it as a child. I just began to appreciate it when I started to learn something about Cape Verdean culture. (Jano)
>
> I didn't learn it. I just watched it many times in my village in the countryside whenever there was a special occasion like a baptism, wedding, or traditional party. I started to learn its history at school. (Isabel)

But for some, the secret of *batuku* was kept even from them and some said they didn't hear or learn much about the *batuku* until after independence from Portugal (1975) or even after that:

> I heard about it, learned it after independence. (Gilda)

Dangerous Women Dancing

Batuku is identified with national pride, resistance, and especially for girls, an indication of their coming of age. The dances and songs, however, are also perceived as "dangerous" because they are seen as erotic and something that emerges from the poorest communities. Julia explains this mix of meanings:

> Well, my father didn't want me to participate in batuku because he thought it to be something erotic for low class people. My mother, who used to dance it as a girl, never motivated me to do so, because she agreed with my father. But I learned it in groups with my friends in parties and festivals. Some people told us it was what every Cape Verdean woman should learn, a sign of Cape Verdeanity. (Julia)

Batuku in Brazil

Portugal carried cultural, religious, and economic ideas and practices to all of its colonies including Cape Verde, Brazil, Angola, São Tomé, and Príncipe, Guinea-Bissau, Mozambique, as well as others. Batuku was a symbol of African identity and resistance to the European colonial influences and particularly the system of slavery they created from the sixteenth century until the late nineteenth century in Brazil. This dance became a problem to the colonial authorities during slavery. Slave masters, police and religious and political authorities argued such batuku celebrations did not contribute to labor output and profits. However, one Jesuit priest said the dance could continue as long as it remained "honest and innocent." But the slaves used their "honest and innocent" dance to reveal hidden cultural expressions of resistance (Reis, 2006).

Captain José Gomes said in 1809 of the *batuku*:

> their bodies are half dressed, they [play] a big tambour, and some of them are decorated with some golden pieces...they continued with their dances not only during the day but continued going a good part of the night, they had a lot to drink... (Reis, 2006)

We can see Gomes was critical of the sensual display of the slaves, especially of the women. This is paradoxical as many white men forced African slaves to have sex or took them as mistresses.

One Priest, Ignacio dos Santos, attempted to prevent the *batuku* during one celebration, by "approaching [the slaves] with apostolic zeal." The slaves defiantly responded by telling the Priest that their masters had seven days a week for entertainment, while they only had one, and asked the Priest to leave (Reis, 2006).

In this context, the slaves acknowledged their master's exploitative position but they did not bow down to it. They did not obey the Priest but instead demanded their time and space for cultural expression. This was especially important, as the Priest stood for the Catholic religion which was "the most powerful instrument of ideological control in Brazilian slavery" (Reis, 2006). The persistence of *batuku* celebrations in spite of the fact that the dance "offended" religious and political leaders provides evidence that Brazilian slaves retained their own political position and did not accept all of the controls placed on them by the masters. Collective celebrations allowed the slaves to acknowledge their oppression and celebrate their values. This was a form of resistance and freedom in the past for the subjects of Portuguese colonialism in South America and in Africa, and Cape Verdean women continue to dance *batuku* as a form of resistance (Reis, 2006).

The contemporary and colonial experience of women dancing in Cape Verde was similar to that of women in Senegal. There, women's dances were a central feature of both life cycle rituals and political meetings. Women were criticized for their dancing by public leaders and their families. Men political leaders in Senegal chided the women for being too suggestive and even banned women from dancing altogether during the rainy season claiming that it might damage the harvest. Sometimes men in their families punished women for dancing (Heath, 1994). Deborah Heath (1994, p. 102) explains in her research on women in Senegal, "[A dance performance] is always a cultural performance...Dance constitutes a form of meaning and action, and like all culture...[it may] provide a vision of power and power relations."

Dance, therefore, is an act that may bring sharp criticism and even punishment because it is seen as a challenge to those in power. Dance may also be acknowledged as an act of defiance by those who are not in power in Senegal and in Cape Verde. "In urban Senegal, when dancers find that those in power have identified their performances as inappropriate, they defiantly embrace their status as other and proceed to dance despite disapproval" (Novak 1990, p. 90).

Identity, Expression, and Resistance

> *Nothing expresses better the collective psychology of people than its music.... The race, the imperial power, the catastrophe of the conquest, the pain of domination, and the richness of the [independence].*
>
> —*(Mendoza, 1998)*

Music and dancing allows men and women to maintain roots in their pasts and a connection to their present. Racism, economic hardships, exploitation, sorrow and violence, unemployment, hopes and dreams, relationships, romantic love, and creative solutions, are all expressed in a culture's music and dancing. Music and dancing preserves the cultural, historical, and oral traditions in a group of people (González, 1999). Music and dancing also construct and shape identity at the same time they serve as forms of expression (Wade, 2002).

Cape Verdean women especially express important feelings and events through dance. They use the cultural expressions that have been deemed appropriately feminine to do that which is not appropriately feminine, challenge authority (Westwood, 1984). The women of Cape Verde are constrained by assumptions about proper behavior and responsibilities but through dance, women release the social constraints that are placed on them (Jensen, 2001). According to the participants, what is not acceptable to express in everyday behavior or in speech, is possible to express through dance. It is one of few opportunities women have to express their joys but mainly their frustrations as well as constructing solutions to the difficulties with which they must contend.

Batuku represents the identity of Cape Verdeans and was a dance of resistance during colonial times. Participants related that it was practiced occasionally by men, but is especially tied to women's expression and even to young women's rites of passage to womanhood. And according to the comment by Julia it is a form of resistance against patriarchy as well. Participants explained that the words in the *batuku* hold deep meanings for women as they sing about problems

Figure 7.1 Rolled up cloth covered with plastic are used as drums in the *Batuku* dance.

with men, famine, drought, lack of jobs, and criticisms against the government. They described *batuku* as an outlet of personal expression for women as they release pain, joy, and sexuality. *Batuku* is a source of hope and strength to counter the tedious and onerous work women do. *Batuku* helps women to cope with problems caused by the difficult social and economic context and provides an outlet to challenge their restrictions. Most interestingly, participants noted in the particular setting of *batuku*, women were freer to voice complaints and discuss problems. In every day speech, such topics would be culturally inappropriate to discuss. *Batuku* is a significant form of resistance by women in the past and today.

Culture is a platform of dissent. This chapter shows women responding to the two poles of injustice, gender formulations that tell them what to be and a long history of colonialism, slavery, and their contemporary expressions in poverty and exploitation. The chapter has also shown women responding in a way that uses an ancient cultural tradition, *batuku* to talk back to those power systems and to reconstruct their own ways of living.

Figure 7.2 *Batuku* is a woman's dance.

Reflections on Chapter 7 Research Methods

Chapter 7 describes an unanticipated way that women resist the pressures and difficulties of their day-to-day life. We made this discovery by first reading and rereading the transcripts many times in order to determine a way to describe our findings and even more importantly a way of analyzing them or understanding them. We coded the transcripts into themes based on nouns and noun phrases: "source of expression," "issues of power," "solidarity under colonialism," and so forth. We then identified an overarching theme that tied the sub-themes together: oppression and resistance. We wrote additional memos to ourselves about oppression and resistance and then in the literature we found that other scholars such as Hurley-Glowa (2006) had observed these same issues in research on *batuku* in Cape Verde. This concept of resistance became one of our lenses, our framework, our *theoretic construct*, for making sense of the data.

We also noted that the resistance was expressed through culture. Throughout our study of Cape Verde, the importance of culture as a vehicle for resistance has emerged. We have included a few examples of poetry and song lyrics that show the way Cape Verdeans have used culture for a very long time to store their history and to express the most essential features of their experience. We heard them in the poems about the sea, the horror of slavery, and the inhumanity of colonialism. In this chapter we add the music and dance of *batuku* as an important repository of the people's history. We also see that *batuku* has served and continues to serve as a platform for dissent. Especially for women, this feminine cultural expression allows the singers and dancers to tell their story. In the next chapter we will identify another form of cultural expression, language, which serves a similar purpose.

In chapter 3 we described this study as a critical research and discussed how that means we went beyond looking at the experiences of our participants in a vacuum to trying to understand them in the context of systems of relationships of power. Harvey (1990) describes this process. He writes,

> [c]ritical social researches take an empty abstract concept and reconstruct it as a historically specific idea which has relevance within a structure of social relations. The reconstructed concept thus goes beyond the particular and is the basis for a critical analysis which reveals the nature of structural relationship [which is] hidden behind the empty abstract concept. (1990, p. 6)

In our research we have taken the concept of dance, particularly *batuku*, and examined what it means for women in Cape Verde. Cape Verdean women live within a structure of social relations that includes a legacy of slavery and colonialism and the contemporary system which includes the power relations between nations of the North and South in a global political and economic system in addition to a system of gender inequality and oppression. This process has allowed us to see that *batuku* is a form of dance but it is much more. It is also an expression against colonial power, slave holders, and today's neocolonial and patriarch representatives of political power. The "empty, abstract concept" of *batuku* has been revealed as a critical platform of resistance.

8

Language as Resistance

Creole is the language that Cape Verdeans use to express what is coming from their soul. It is the means to say exactly how you feel about everything. If people who come here don't understand Creole it would be difficult or almost impossible to know Cape Verdean's culture. In another language, it is difficult to find the words, the expressions that we exactly need.

—*(Carla)*

In this chapter, we examine the perceptions Cape Verdeans attach to the importance of Creole in order to fully understand the culture. In addition, we listen to Cape Verdeans speak of important problems in their lives and explore the ways they use Creole to both reflect and challenge the power relations by gender and class in Cape Verdean society.

Creole of Cape Verde has been a means of communication shunned and disparaged by the powerful, but preserved and celebrated by the population who speak the language on a daily basis among themselves. Published literature in Creole was forbidden during Portuguese colonization until shortly before Cape Verde's independence in 1975. Portuguese officials also prohibited spoken Creole in government buildings (Lobban, 1998; Ludtke, 1989). Even today, Cape Verdeans learn Portuguese in primary and secondary school and spoken Creole is not allowed inside the schools, though Creole is Cape Verde's first language among nearly all of its citizens (Lobban, 1998). Even though Creole was prohibited for a very long time during colonization, the language has been published in a considerable amount of music, literature, and poetry, including the writings of well-known poets Jorge Barbosa, Eugénio Tavares, and Baltasar Lopes. Creole persists as a significant linguistic feature not only for the people who live in Cape Verde but also for the Cape Verdean identity that has spread as Cape Verdeans have emigrated to the United States and other parts of the world (Gonçalves, n.d.). This chapter begins with three sections that describe the political character of language; stereotypes of language use; and the origin and development of pidgin and creole languages (Carter & Aulette, 2009).

Identity, Contact, and Power

Language is a fundamental part of our character, an expression and a mirror of what and who we are. Ethnic groups in particular use language as one of their identifying features. But just as all ethnic groups are not equal, all languages are not equal either. The power inequalities that exist between different sections of a society are reflected and reproduced by language (Carter & Aulette, 2009).

Language represents and creates relationships between the speaker and the listener and these relationships can include the factors of power, control, and domination. When ethnic groups recognize each other as dissimilar in terms of power relationships, language boundaries resemble clear-cut borders that are protected (Wright et al., 2001). Language contact is contact between people speaking two different languages. Language contact situations are comprised of two sometimes conflicting goals—the need to communicate and interact efficiently and the need to safeguard group identity (Winford, n.d.; Carter & Aulette, 2009).

Cape Verde represents a diglossic speech community, meaning "society-wide" use of two linguistic varieties. In Cape Verde the two languages are Portuguese, which is a "high" language, and Creole, which is a "low" language (da luz Goncalves, 2001). Ferguson's (1959) classical description of diglossia was divided into two categories: a "high" language H, used for education, literature and formal purposes, and a "low" language L, used for informal purposes (Mendoza-Denton, 1999). Speakers of "low" languages challenge the power of those speaking "high" languages. Thus the use and choice of a language is never socioculturally or politically impartial (Murray, 1998).

Misconceptions of Language and Culture

According to Bradley (2002), attitudes about language connect to attitudes about the culture itself. Further, we regard a person's character and background based on their language (Wolfram, n.d.). Such prejudices create much pressure for speakers of disfavored dialects to abandon their native speech for some approved pattern. The political forces behind the decision to abandon one language for another are sometimes hidden by rationales of efficiency. Sociolinguists observe that the abandonment of language among immigrant communities does not occur from a simple change within the home, but instead, abandonment occurs from a change in attitude about the language's status. Brenzinger et al. (1991, p. 38) writes,

> The decision to abandon one's own language always derives from a change in the self-esteem of the speech community. In the cases of language shift one could observe that members, very often the younger generation of minorities, regard their own community as being inferior. Those members frequently try to change their "negative" social identity by adopting the language (and social identity) of the dominant group....Language shift thus has to be understood as one possible strategy for members of minority groups who have developed a "negative" social identity to change their inferior position. In cases where this strategy is chosen by all members of a minority speech community we could expect the extinction of the old vernacular.

Einar Haugen argues against any form of abandonment of language. He argues that this process of extinction of language is a dreadful event in the human experience. He maintains that when languages disappear we are witnessing a kind of cultural genocide. He says,

> and yet, who are we to call for linguistic genocide in the name of efficiency? Let us recall that although a language is a tool and an instrument of communication, that is not all it is. A language is also a part of one's personality, a form of behavior that has its roots in our earliest experience. Whether it is a so-called rural or ghetto dialect, or a peasant language, or a "primitive" idiom, it fulfills exactly the same needs and performs the same services in the daily lives of its speakers as does the most advanced [*sic*] language of culture.... (Clark et al., 1998)

As Brenzinger (1991) argues, not only languages are threatened when native languages are discarded in favor of "better" languages. Ways of being, the identities of individuals and communities are also "made to disappear." This process, however, is complex and can also work in the reverse as languages of colonial powers are used and altered to create pidgins and then creole languages. When speakers of different languages first come in contact with each other, at first evolves some form of auxiliary contact—a language native to none of them which are known as pidgins. Eventually these pidgins become the creole languages which exist today (Bickerton, 1981; Carter & Aulette, 2009).

Pidgin and Creole

> *Pidgins typically result from traumatic contacts between various ethnocultural groups, in which one group assumes socio-political dominance over the other(s). This sociolinguistic clash typically occurs in the course of events such as invasion, slavery, or other types of migration. Creoles develop over time when displaced populations are forced to remain...or when an occupied territory remains under the tutelage of an invading force (...many parts of Africa). A new language variety has to be fully developed to mediate communication or—to use a common definition of creoles—the creole emerges when the preliminary pidgin code acquires native speakers.*
>
> —*(Escure, 2001: 55)*

This process began to unfold on the West African coast in the fifteenth century. The first contact between Portuguese sailors and West Africans probably took place in the mid-1400s. Portuguese colonization of the West African coast led to the development of a new language (*The Papiamentu Language*, n.d.). Soon thereafter, Portugal joined in the slave trade, forcing millions of West Africans into slavery until the nineteenth century in which the colonial language clashed with local language. Communication among slaves was complicated as West Africa was one of the most linguistically diverse areas of the world (Mullin, 1994). To communicate with each other and their Portuguese owners, slaves began to develop and speak in coastal Creole during the long stays in West African harbors, until their passage across the Atlantic. Linguistic features of the language of Mande, Fula, Balante, and other ethnic groups from the Upper Guinea coast mixed with Portuguese, and over time formed Creole. This may have become a secret language shared by the slaves, beyond the understanding of the slave owners (*The Papiamentu Language*, n.d.; Coleman & Daniels, 2000).

Today there are Portuguese-based creoles, English-based creoles, French-based creoles, and Dutch-based creoles. Creoles were and are exceptionally distinctive because their formations led to new communities of speakers. Creoles developed a personality of their own, along with new rules of usage, which defined the group identity of its speakers (Carter & Aulette, 2009).

Cape Verdean Creole

Dating from the fifteenth century, Creole of Cape Verde, also written as Kriolu, Crioulo, Papiamentu, Kabuverdianu, or Caboverdiano, may be the oldest of the many different Creole languages spoken today (Gonçalves, n.d.). Continental Portuguese is the language of business, education, government, and literature while Creole is the language of daily life in Cape Verde, meant for family and informal situations, including jokes, songs, stories, proverbs, and feelings and emotions (Ludtke, 1989). Speaking one language for certain activities and a second language for others is not unique to Cape Verde. Residents of Luxemburg speak French in school, German for reading newspapers and a local German dialect, Luxemburgish, in the house. Similarly, Paraguayans speak Spanish when doing business but tell jokes in the local indigenous language, Guarani (Bryson, 1990).

Cape Verdean islands are divided into Barlaventos (Windward Islands) or the Northern Islands and Sotaventos (Leeward Islands) or the Southern Islands. Differences of dialect are found between the Northern and Southern Islands. Creole in the Northern Islands is said to be more "Portuguized," reflecting a larger influence from the official language of Portuguese. Creole in the Southern Islands is said to be more "African" and less Portuguized. The Southern Islands also retain more of their African traditions even beyond language with a lesser degree of European influence.

Keeping a distance between Creole and Portuguese was used by the Portuguese colonizers as a way of maintaining social distance. Creole was thought of as a low-status dialect and a sign of inferiority. Creole was not to be used by educated persons, and was argued to be a barrier to theoretical thought and educational progress. Knowledge of Portuguese was presented as a better language and a requirement for jobs other than manual labor. Among many intellectuals, however, spoken and written Creole was a sign of resistance toward Portuguese rule (Gonçalves, n.d.). The use of Creole was a significant part of the anti-colonialism struggle especially for Amílcal Cabral, an intellectual and political leader of the movement, and his followers in the struggle for national liberation of Portuguese colonies in the mid twentieth century (Cabral, 1970).

Until recently Creole existed only orally while scholarly work on the language was inhibited because study by an outsider was almost impossible without written texts. Since independence of Cape Verde in 1975, some school texts have been written and published in Creole. Cape Verdean literature is now gaining much popularity (Ludtke, 1989).

Today, some government officials have argued for the adoption of Creole as the nation's official language. This has been opposed however with opponents arguing that communication with the outside world could not be conducted efficiently in Creole. Even so, the emotional meanings of Creole are so vivid and lively, its essential place in the national consciousness can be easily seen (Ludtke, 1989).

Popular literature and *mornas*, a slow piece of music played on stringed instruments describe emotions such as *sodade*—longing. Longing for a loved one, longing to return to the islands, longing to return to better times. The imagery used in the mornas and other forms of poetry was one of the ways Cape Verdeans preserved their ethnic consciousness during the years of repression (Ludtke, 1989). Poets Jorge Barbosa, Eugénio Tavares, and Baltasar Lopes especially wrote about the difficulties of life on the islands such as isolation, migration, famine, and drought. Eugénio Tavares wrote the first book with Creole text called *Mornas—cantigas crioulas*. The beauty of Creole in poetry and music contradicts the idea that Creole prevented artistic creativity.

Today Creole is the most widespread and clear symbol of Cape Verdean identity within the country and around the world (Lobban, 1998). The language links more than 700,000 Cape Verdeans in the diaspora—Germany, Italy, Luxembourg, Netherlands, Spain, Portugal, Angola, Senegal, Brazil, France and the United States. Manuel da Luz Gonçalves writes, "from California to Boston, [Creole] is part of our identity, our way of knowing, but also often our access to the world through radio, TV, and the educational system" (Gonçalves, n.d.; *Languages of Cape Verde Islands,* n.d.; Carter & Aulette, 2009).

She Is One of Us

Though Creole is Cape Verde's first language among the majority of Cape Verdeans, Portuguese is associated with the upper class in Cape Verde society. If an outsider to Cape Verde only speaks Portuguese they would not belong to the in-group, but they would be perceived to have a higher status. Instead of being seen as Cape Verdeans, they would be distinguished as strangers, aliens, and outsiders. They would miss the reality of Cape Verde, the daily jokes, and proverbs. But they would also be perceived to be in the dominant group.

As an outsider living in Cape Verde for three years, Carter often thought about the importance of Creole and its contrast with Portuguese. She knew that Creole was a significant aspect of Cape Verdean culture and wanted to be integrated into that culture as much as she could. She also knew, however, that in spite of her best efforts to become an insider, an equal to her friends, colleagues, and neighbors, she would never be Cape Verdean. Her own ways of being and ways of knowing would always be imposing. She could however, reduce the barriers by learning and speaking Creole.

When Carter met new friends in Cape Verde, she was often looked at with apprehension. Their faces seemed to say, "who is this white woman and what does she want?" If she began talking and making jokes in Creole, however, the barrier was broken. Her imposing status of foreigner was diminished. They let out their breath, smiled and laughed. They could identify *with* her. Many times she heard, "*ele é de nos*." She is one of us.

She visited many homes whose walls were made of cardboard or stuffed with plastic bags or families that lived on the side of a public dump. She entered the homes and spoke their local tongue. In either situation of meeting new friends or visiting the homes of those she already new, if she spoke Portuguese, rather than Creole, she would be seen as an intruder, an outsider. Cape Verdeans would feel less comfortable talking with her. According to participants, when outsiders, who do not speak Creole, come to Cape Verde they present a special problem for Cape Verdeans who want them to discover and understand the true character of the country. The following section focuses on the perception of that problem from the point of view of Creole speaking Cape Verdeans (Carter & Aulette, 2009).

Gateways to Cape Verde

At the beginning of this chapter, Carla tells us how important, uniquely essential, knowledge of Creole is to understanding Cape Verde and Cape Verdeans. She provided us this reflection in response to the question we asked of all the student participants: "Some people say that if a person does not know Creole, they do not know Cape

Verde. Do you agree with this statement? Explain why you do or do not agree with this statement." (See appendix C for complete list of questions.) When participants wrote about their perceptions of the importance of language in Cape Verde, their observations fell into three types. One group saw Creole as a gateway that must be entered in order to have any understanding of the culture and people. Another group suggested that Creole was a key but it wasn't enough. Knowing Creole was important but other factors were also essential. A third group believed that Creole could be an important gateway but it was not essential and other gates could work just as well.

Leigh was one of the participants who fit into the first category. She argues that Creole is an essential tool, not just a gateway, but a "weapon," for being able to absorb Cape Verdean culture and for one to be absorbed into Cape Verdean culture. Her perception is that it is a challenge to find access to Cape Verde and Creole is the essential "weapon" to find one's way to the inside:

> Creole is our mother tongue and to know Cape Verde you have to speak Creole. Language is a weapon used to integrate in a society. If you don't know the language, you can't integrate. (Leigh)

Others said that understanding Creole is important for grasping songs and jokes, and also for expressing intimate feelings. They felt finding equivalents in Portuguese are impossible because Portuguese cannot reflect the Cape Verde experience. Lacking knowledge of how to speak Creole creates an insurmountable barrier to understanding Cape Verdean culture and reality:

> Someone who doesn't understand Creole will miss the meaning of songs and jokes, which reflect the Cape Verdean reality. Moreover some words are impossible to translate into Portuguese. (Deusa)

A second category of responses to the question of the importance and role of Creole to grasping Cape Verde, was expressed by participants who argued that the language was important but that it wasn't the only critical factor. Mark explains that language is important but to know a country well, time spent with the people is equally or even more important.

> Language belongs to our culture, and to know a culture of a country or a people, one should know their language. It is the main part of a culture; it is the way to transmit habits and customs, to know better

> the people. I think to know better a country, you should spend a lot of time with its people. (Mark)

Although it is not the only vehicle for understanding Cape Verde, Creole was asserted as an essential factor. Both insiders and outsiders are prevented a deep knowledge of the people and their culture and history if they cannot hear about the language of the people, which itself reflects the history of Europeans and Africans colliding to form Creole. A third category asserted that Creole is important but learning the language is not essential for integration into society. They felt that language was one way, but not the only way, of understanding Cape Verde. Claudio explained,

> Creole language is just one element of Cape Verdean culture. What identifies Cape Verde and Cape Verdeans are the language, the climate, the Catholic religion, music and dance (funana, batuku, morna, coladera, etc.), the food made by corn (cachupa, camoca, xerem, etc.), the peaceful country, the friendship and hospitality of its people, etc. If one doesn't know the language, one can know other elements of the culture. (Claudio).

Francisco further explains that this option is only open to wealthy people since finding out about Cape Verde through these other means requires travel, which is too expensive for the poor. Francisco is focusing on the ability of Cape Verdeans, rather than outsiders or tourists to gain knowledge about the nation. He argues that wealthy people, who do not speak Creole in the home, could understand Cape Verde by traveling to the different islands. Travel however is reserved for the few, for the upper class of society. Travel and seeing all of Cape Verde as well as studying about the country, according to Francisco, however, might be even more important than knowing Creole. He explained,

> There are a lot of local people that belong to the upper class, the language they speak at home with their families and at school is Portuguese and I think they know Cape Verde very well because they have money to go and see the other islands to practice tourism and so on. For those only speaking Creole, they might know Cape Verde superficially. The history of Cape Verde, in schools, is taught and given in Portuguese. (Francisco)

Some participants also felt learning Creole was not essential for outsiders to understand Cape Verdean society, except in the case of communication with Cape Verdeans. They believe that outsiders who do not know Creole can come to Cape Verde and understand the culture, even though they can't talk to people or make friends. Knowing the language would help an outsider to make friends who would help them to understand the culture but language was not essential. Mario explained,

> We don't know the reality/knowledge of one country only through his native language, where for example we have in Cape Verde a lot of tourists that do not know our language but they know so many things about our country. Their only problem that they can have if they don't know Creole is communication with Cape Verdeans, but in Cape Verde, we have some people that help the tourists learn Creole in a fast way. (Mario)

The participants disagreed about the importance of language. Some argued that knowing Creole was essential while others said it was less important (Lucy, 1997). All of the participants, however, believed that language could play an important role in allowing to or prohibiting people from understanding Cape Verde and its people (Carter & Aulette, 2009).

What Might They Miss?

Clark et al. (1998) writes that to know a language is to speak and be understood by others who also know that language. Further, every human language has been created by and modified to meet the needs of its speakers and every language has important areas that are more in depth and dynamic than others. We asked participants to explain what they thought someone would miss if they did not speak Creole. They described three essential roles that language plays. They explained that language allowed people to communicate their wishes, to socially interact. Second, they described the ways that language not only establishes fleeting connections, but helps to build bridges among people, creating social relationships. Third, language both reflects relationships of power and allows one to challenge lines of power.

Participants were asked: "If someone only speaks Portuguese in Cape Verde, what exactly might they miss? Try to think of a specific example of something a person would miss if they only spoke Portuguese."

Participants identified several problems that might occur if an outsider did not speak Creole, including practical situations like being overcharged at the market or unable to communicate at all with rural Cape Verdeans who only speak Creole. They also mentioned those who do not speak Creole risk social exclusion. Language represents a tool for achieving certain tasks with others such as negotiating a sale, but language also represents a social link between people. Mark explains how language works to build bridges among people:

> I think what is missing is a connection between the two people, for example when someone comes to Cape Verde, he is new in the country and he feels a little strange, so in a foreign country the adaptation can be very difficult, the people have to be close to know each other, and that connection is language, Creole. (Mark)

Mario further explains the social connection provided by language when he goes on to say that if someone cannot speak Creole, he or she will become isolated with few friends:

> If a person only speaks Portuguese in Cape Verde, that means that this person can only be Portuguese or Luso-African, because it is difficult to find people in Cape Verde that only speak Portuguese. If a person insists on only speaking Portuguese in Cape Verde, that means this person has few friends, and has problems in relationships with people in the case if he's new in the country. (Mario)

In contrast, however, Liz has a less stigmatized view of Creole but she too believes that not being able to communicate with Cape Verdeans who speak only Creole would mean that one would miss a big piece of the picture. The inability to communicate would create barriers to being close to others, learning about the country, and finding social connections.

> They would miss the opportunity to learn a beautiful language, Creole, and they also miss the opportunity to make many Cape Verdean friends. (Liz)

Others argued that the loss would be especially significant because the heart of Cape Verde is among those people who live in the countryside and are less influenced by outsiders and less likely to speak Portuguese, relying on Creole alone for communication:

> The Portuguese language, in spite of being the official language in Cape Verde, is not spoken by the majority of Cape Verdeans. Most of

> those who can't speak Portuguese live in the countryside, and that's where the influence coming from abroad has been felt the least. This means that if someone wants to know the real Cape Verdean culture, he/she should go and visit the countryside. And if he/she can only speak Portuguese, he/she will have problems in knowing our culture. (Dani)

Vanda explains that not knowing Creole will close other doors preventing people from learning about other, essential features of the culture:

> For example, if he/she went to the countryside Portuguese cannot be understood at all, it is difficult for him/her to deal with people there. If he/she needs to know the culture, like the history of *batuku*, for example Cape Verdeans will not feel easy/comfortable to talk with them in Portuguese. (Vanda)

Sara points out the social link created by language are not just generic connections but include aspects of power. She describes the problem of being over charged at the market, and asserts that the problem is not just practical but reflects political tensions. She believes language reflects social distance and social class. Usually this means that those with the most power speak the most respected language. In this case, however, the roles are reversed and those in non-powerful positions of being sellers in the market have the upper hand:

> Since Creole is our native language and many people don't speak Portuguese here, I think that if someone only speaks Portuguese in Cape Verde, he/she will have difficulties to be understood, for example in the market. They might be overcharged because the sellers will think that the person is rich. (Sara)

Alberto elaborates on the political relationships embedded in communicating (or being unable to communicate) in Creole. He argues that knowing Creole is especially important when interacting with rural Cape Verdeans or illiterate Cape Verdeans. Like Dani, Alberto notes that not knowing Creole inhibits one's ability to know about people in rural areas. But Alberto explains the power differences of people in those areas and the political lessons about Cape Verde that would be undiscovered if one were not able to speak Creole. Alberto identifies non-Portuguese speaking people as those who are of a lower status than those who do speak Portuguese.

> If someone only speaks Portuguese perhaps he/she would have problems to communicate with illiterate/uneducated people, because there

> are some people who have problems in understanding and speaking Portuguese. (Alberto)

According to the participants, people who speak only Portuguese would be placed in the out-group, although they would also be perceived to be of a "higher" class. But their status as above Creole speakers and their inability to communicate in Creole would cause them to be foreign and kept at a distance.

> They can't know deeply Cape Verdean culture, because there are some sayings, proverbs, jokes, anecdotes that are funny only in Creole. If someone only speaks Portuguese, they can be out of the society or group, or they seem to have higher status or pretend in having. (Claudio)

> The people who only speak Portuguese in Cape Verde—they are seen as foreigners and even the Cape Verdeans who speak only Portuguese think they are important/high people, because of the language they speak. So they miss the contact of Cape Verdean reality and of course they are far from Cape Verdean language or the daily conversation. (Jose)

Participants believed that knowing Creole is at least useful and often essential to understanding the country and its people. They maintain that a non-Creole speaker cannot integrate, make friends and will face practical problems such as being charged higher prices in the market and often will not be able to communicate at all with Cape Verdeans who speak only Creole. According to the participants however, Creole is also a social class indicator and those who do not speak the language may be perceived as "higher" class, as outsiders, and someone from which to maintain a social distance.

Language reflects a complex system of political relationships among those who speak Creole and those who do not. The gendered character of these political relationships emerged in the second data set when we interviewed women about their lives in Cape Verde. (See appendix D for complete list of questions.) Women used colorful proverbs and sayings in Creole, to express their opinions, acknowledge their difficulties and suggest sources of those problems such as colonialism or patriarchy (Carter & Aulette, 2007). The next section looks at how Creole proverbs are used by women.

Proverbs, Sayings, and Gendered Power in Creole

Proverbs are regarded as traditional wisdom in a culture. They are reflections of a culture and summarize universal truths within that culture

(Ntshinga, 1996). A special regard is given to those who use proverbs properly; the respected nature of the proverb urges the listener to be in agreement with the speaker (Hewer, 2000–2001; Ntshinga, 1996).

Sayings reveal more of a local character than proverbs. Sayings originate as colorful, impromptu remarks which may later become a fixed saying (Hewer, 2000–2001). They are simple expressions of wisdom or truth (Randall, 1991). Both proverbs and sayings reflect a culture's norms and values (Ntshinga, 1996).

Several scholars have found that women and men differ in their use of language, including the ways they use proverbs and sayings. Women in American society for example have a language of their own, one different from and perceived as inferior to the language of men. Women's use of language is often not as aggressive, confident, authoritative, or direct as men's language. This doesn't mean that women are naturally prone to speaking in this manner. It means that the social construction of femininity includes certain ways to use language. Women are expected to speak in less confident and authoritative ways and if they do not, they may be perceived as harsh or unfeminine. Scholars agree that women's language is symbolic of women's lack of power in society (Ntshinga, 1996; Coates, 1998; Thorne et al., 1983; Wodak & Benke, 1997).

Cape Verdean women's use of Creole language reveals this link between power and gender noted in the literature on American culture. We found that language use mirrored the lower status of women in Cape Verdean society and their inability to often speak directly about the problems they face with poverty and non-monogamous men. In addition, our observations reflect arguments about gendered power, summarized by Holmes (1998) who writes that women speak in ways that create solidarity, while men speak in ways that reinforce their authority, control and power. Women's use of Creole illustrates a social construction of femininity in Cape Verde in which women are not powerful. At the same time, however, Creole, especially Creole proverbs and sayings, are tools women use to challenge their less powerful position and the individuals and social systems that have repressed and oppressed them.

Many scholars have noted the misogynist character of proverbs in a range of cultures. For example, proverbs often construe women as shrewish or ignorant (Escure, 2001; Maurice, 2001; Braun, 2001). In Cape Verde, however, women use proverbs as tools for exposing and challenging gender inequality and gender injustice. Women draw on proverbs and sayings to illustrate what they are not getting out of a relationship, work situation or the society as a whole. They are a critical response to her life experience and world view (Ntshinga, 1996). Ntshinga (1996, p. 13) writes,

> Communities do not listen to expressions of African women's experience. But irritation, anger and impatience within groups of women have resulted in [women] coining new proverbial expressions of emancipatory concerns and ideas. These expressions are characterized by frankness because women have now come to the realization, more than ever before, that their world will be destroyed unless they do something urgently to save it.

In our research when women in the community were interviewed, they used colorful Creole proverbs and sayings to describe, explain, and challenge their situation. These interviews were not originally intended to explore the issue of language. We did not ask the women what they thought about Creole or how they used the language. The role and character of language use in describing important problems, however, emerged as a dominant theme in our analysis. The women used Creole proverbs to describe and illustrate their frustrations and their strategies for coping with or confronting the difficulties they faced such as relations with men, infidelity, lack of child support from fathers and work situations. Women were asked, "Do the men in your life live up to your expectations?" Women used sayings to explain they did not have confidence in men, they believe men are dishonest, and that men are constantly involved in illicit sexual affairs. Rita, a primary school teacher and a single mother of four children said, "*omi pa mi e so makaku.*" [Men are monkeys] and "*omi largam ku quatru fidju na pe pam kria.*" [My husband left me with four children around my foot]. Carla, who had previously taken her husband to court because he was not helping to provide for her and their seven children, said, "*korda ta rebenta na mas fraku.*" [The weakest strand is broken / men are like the weakest link].

Rita and Carla's comments use colorful sayings and proverbs to quickly convey highly critical, even offensive ideas about men: they are weak and animal-like, beneath humans. But the sayings allow them to do it in relatively light-hearted, even humorous manner. The sayings and proverbs also allow them to convey deeper meanings in a limited amount of words. Rita's comment on being left a single mother with children tied around her feet makes it easy for the listener to imagine the difficulties she feels in this task and the blame she places on her husband for "tying her up" in this way.

Lota, a street vendor whose husband was physically abusive, used two contradictory sayings to describe the change in her husband over time, "*the first year of our marriage we lived in the seventh heaven of delight... but as it is said, there is no rose without a thorn.*" Once again the sayings convey much more meaning than the limited number of words might imply. We can imagine the delight she first found in her husband and the pain he caused her later in their marriage.

Women also used sayings to describe their own feelings when men let them down by having affairs with other women. Women were asked, "How would you feel if your partner began a relationship with another woman?" In chapter 3, we heard Pauline say, "*n ta qui marra corda na barriga pam aguenta.*" [I would tighten my stomach with a tie and bear it]. In her use of this saying, Pauline is able to express her strong feelings. Her pain does not remain invisible. She is able to make the powerful assertion that men who cheat on their partners are causing as much trouble as if they had caused a death.

Women not only expressed their sorrow and resentment, they also expressed their pride in being able to find solutions to the problems they faced as a result of undependable men. When asked about what she thought about her work situation, Helena, a street vendor with five children and an estranged husband, said, "*sacu basiu ca ta sakedu.*" [An empty bag can't stand up]. Helena believed someone with an empty stomach couldn't stand on their feet but may faint and fall. Liking work or not was irrelevant. Helena does not directly criticize her husband or challenge the power he has over her life but the proverb she uses quickly communicates the idea of how significant his action has been—leaving her a single mother is like leaving her an empty bag. It also tells us, however, that she can and did rise to the occasion making the bag—and herself—stand up.

All of these sayings show how Creole proverbs can be used to quickly make a point. They present difficult and sensitive subjects in ways that are "prefab" allowing the speaker some distance from the topics but also allowing them expression of strong opinions. They reflect a power relationship in which women are not powerful. They also, however, represent a challenge to those power relations since the women's ideas are expressed from their own critical point of view. Proverbs and sayings express solidarity, provide support, and show criticism of power structures in their relationships and in the broader society's systems of gender inequality (Carter & Aulette, 2009).

Creole in Cape Verde

The participants in our study said that it was essential to know how to speak Creole in order to integrate into Cape Verde. They reported that if individuals spoke Portuguese, rather than Creole, they would be perceived to be of a higher class but they would be placed in the out-group. Creole might be seen as a marker of lesser status. For the Creole speakers we interviewed, however, it was also a sign of solidarity and a way to distinguish oneself from those on the outside who spoke only Portuguese. Outsiderness cannot be erased by learning Creole, but the social barriers between Cape Verdeans and those who wish to be their allies are greatly reduced by speaking Creole.

Creole language and its proverbs are especially likely to serve as a source of expression for women. The colorful proverbs and humorous sayings in Creole have hidden meanings and allow women to stand up for themselves, to have their say, and thus can be described as a way of coping and as a way of meeting challenges in a difficult economic and social context as Cape Verde.

Colorful proverbs, humor, and irony offer ways to express ideas by individuals whose social conditions are limited. Humor, irony, and indirect communication can be quite powerful in some cultures (Stewart & Bennett, 1991). Westerners, accustomed to directness, sometimes neglect to see the value of indirectness. When women talk about sensitive issues such as poor relations with men, lack of child support and lack of work, directness may not be the best policy. Women instead express these issues through humorous proverbs and sayings (Carter & Aulette, 2009).

Reflections on Chapter 8 Research Methods

Chapter 8 looks at the importance of speaking Creole, especially for outsiders who wish to understand Cape Verde and Cape Verdeans. Like the previous chapter on dance, this chapter also illustrates another unanticipated way women resist the difficulties in their lives. The chapter is divided into two sections. These two sections report two sets of data. The first set of data is from the information we obtained from the student participants. The second set of data comes from interviews the student assistants conducted with women in Cape Verde. The report of the first data set provides us with a context for understanding the importance and use of language, particularly Creole in Cape Verde. The second set of data tells us how language is used every day as a vehicle for challenging the social construction of femininity.

The analysis of the data was done differently with the two data sets. In the first set of data we did a straightforward thematic review. We looked at the themes that emerged in the discussion of the importance of language in Cape Verde from the point of view of the students. In the second set of data we move the analysis up a level by using the tool of critical discourse analysis. When we reviewed the transcripts of the women talking about the men in their lives we noticed that they used language in a particular way. The manner in which they answered the questions, the words and sayings they used were even more revealing than the answers themselves.

The use of proverbs and sayings were a way of structuring the discourse that revealed the underlying political relationships of the women to the men they were describing and to their society itself. Their use of language also revealed a way to challenge those political relationships. The women were able to express themselves honestly and forthrightly but in a less blunt or confrontational way.

9

Conclusions: Everyday Acts of Resistance

The deep elemental stirrings that lead to social change, begin within the hearts of men [and women] whose thoughts have hitherto not been articulate or who have never gained a hearing or whose needs are therefore ignored, suppressed and treated as if they did not exist. There is no revolution without a voice.

—*(Merton, 1980, p. 72)*

We initiated this study by asking how globalization has shaped women's personal stories, and by examining the ways women are resisting the challenges presented to them in a globalized world. As we began to explore these questions, we were thinking of globalization as mostly an economic process and we were looking for ways that globalization might open up new economic opportunities but especially how it created more economic oppression of women in the South. We also started with a belief that this economic oppression would elicit political dissent in the form of protest and calls for policy reform internationally and within nation states.

We learned that this conceptualization of both oppression and resistance is only part of the story. Women in Cape Verde are facing serious economic difficulties as globalization engulfs their nation. They are also responding by organizing forums for education and dissent and are promoting policy change or policy implementation to support them in their struggles to survive. We found, however, that both our conceptualization of oppression and our conceptualization of resistance fell short of the fuller picture women in Cape Verde present of the difficulties they face and the challenges they make to the existing relations of power.

By listening to the voice of Cape Verdeans we learned that women not only must deal with the economic consequences of globalization, they also face the difficulties presented to them by the systems of masculine dominance that are also central to the power structures in their lives. Both economic globalization and systems of gender inequality profoundly affect their lives and these two are intertwined in complex ways. We also learned that resistance not only takes the form of organized political dissent, it also is an aspect of every day life, especially the everyday expression of culture through language and dance. In this last chapter, we explore the theoretical connections among all of these issues. We also consider the theoretical and practical implications of broadening our vision of oppression and resistance in order to find all of the sources of energy, creativity, and power to make change that will create better lives for all of us.

Women in Development

In the 1960s, the world began to change as nations broke the old bonds of colonialism. Most leaders of both old colonial powers and newly independent nations believed the answer to reducing poverty faced by the newly emerging independent countries of the Global South was the "trickle down" approach. This idea included wealthier nations giving aid to the governments of poorer ones and assuming it would be dispersed among all citizens, including the poor. Trickle down theorists argued that development meant building economies that were ruled by a few and benefited those at the top but that their wealth would make its way down to the rest of the people spreading the world's wealth more evenly not only to all nations, but to all classes, races, and genders within those nations.

Ester Boserup (1970) challenged this point of view arguing that many of these so-called development projects overlooked women and in fact, weakened women's economic position. Development projects, for example, provided men training in technological skills while women were ignored or excluded. Aid programs, for instance, might ship in tractors for agricultural work but only train men how to drive them when women had been half of the farm labor before the tractors arrived (Parpart et al., 2000). As recently as 2008, the UN Food and Agriculture Organization in Kenya noted that although women produce 80 percent of food in Africa, traditional farm support still winds up in the hands of men. For example, although women are often farmers, banks distribute low-interest agriculture loans mostly to men, because the men have title deeds for collateral. Government training and tools are also provided to men resulting in ironic and wasteful situations such as farm tools being provided that are too heavy or too long for women to use and therefore lay unused in fallow fields (Alsop, 2008). These policies continue to be challenged all over the world. In the specific case in Kenya, for example, organizations such as Wakulima Women's Group and the Molembe Women's Group have successfully created Rural Outreach Programs that support women farmers (Alsop, 2008).

Since Boserup published her insights, feminist scholars have offered three more models for understanding the global economy and creating policies to try to alleviate the economic difficulties and to eliminate the gender gap around the world (Rai, 2002; Smith et al., 1998; Levy, 1996; Moser, 1993; Razvani & Miller, 1995). The first of these is called WID, *Women in Development;* the second is WAD, *Women and Development*; and the third is GAD, *Gender and Development.*

The WID approach is similar to Boserup's because it accepts the general framework of development theorists except that it pays attention to women. WID maintains that poor nations should model themselves after wealthier ones and that the women in all countries, but especially those in the Global South, should become more like men by moving into the paid labor force to work alongside men. International economic policies should facilitate these transitions by allowing multinationals to build factories and invest in agribusiness while those corporations should be sure to include women in their workforce.

In its recognition of women, the WID approach is a welcome step forward, but it has several weaknesses. First, based on their belief in a modernization prototype, WID advocates maintain that nations of the Global South would and must develop by following along the path of Western values, economies, and technologies. The WID approach: assumes that Western governments have the correct solutions; assumes that Western governments have the only solutions; and it overlooks the input of citizens from underdeveloped countries (Parpart et al., 2000).

Second, WID solutions ignore the fact that although there is a gendered division of labor in every society, women are active and productive workers all over the world. WID's assertion that women should be integrated into the economies of the Global South implies that they are outsiders, but women are already in the labor force. Many women earn wages in factories, fields, and offices. In addition, women may not be receiving wages but are trading and selling goods in the market or working in family businesses especially on farms or in other small businesses. Furthermore women are working to raise children, as well as to feed, clothe, and provide health care for their families. In fact, when both unpaid and paid work is considered, 66 percent of the world's work hours are by women (UNESCO, 2007).

As these weaknesses were pointed out by feminist scholars and activists, their critique developed into another framework to replace WID. This perspective, known as *Women and Development* (WAD), argues development plans needed to make visible the labor, paid and unpaid, that women do and the importance of their work. Rather than calling for integrating women into the labor force, WAD maintains that plans need to recognize the ways women are already integrated, in fact, integral to the labor force all over the world. The problems women face as members of these productive forces need to be addressed. WAD advocates believe that international economic policies should recognize, support, and compensate the invisible work of women.

WAD scholars also assert that nations of the Global South need not and perhaps should not follow the Western model in order to improve their economic and social standing. WAD advocates maintain that each nation's history, resources, and culture need to be incorporated into plans for development. The WAD perspective questions whether the model of the Western developed nation is the only or best way to build effective economies. WAD theorists argue that a range of possible paths to economic development should be created in order to avoid some of the pitfalls of Western capitalism, such as environmental degradation, huge gaps between the rich and poor, and inequities between women and men.

The WAD approach also has its limitations, however. It sees women as a monolithic group and disregards divisions among women by critical differences such as social class and race ethnicity. In the 1980s, some feminists and development agencies began to criticize both the WID and WAD approaches and created the term GAD, *gender and development*.

This third approach criticizes WID and WAD for treating women as a homogenous group (Moser & Moser, 2005). GAD scholars and activists point to the enormous differences in women's experience and call for a conceptual model and policies that recognize the diversity among women. The GAD approach asserts that women's oppression is experienced differently based on women's race ethnicity, social class, and culture, as well as their respective country's colonial history and position in the current global economy (Parpart et al., 2000).

GAD also sees gender equity as part of a larger exercise of human rights and claims a mission of transforming the global economy to one that is based on human needs and cooperation rather than profitability and competition. The goal of the GAD perspective is gender equity, as well as equity in regard to many other social cleavages (Parpart et al., 2000).

This book represents the GAD approach. In particular we have emphasized the unique experience of women in a specific place, Cape Verde, with a distinct history of slavery, colonialism, and globalization. We also have focused on low-income women who are struggling to keep their heads above water as global debt, Structural Adjustment Programs, poverty, migration, and violence wash over them. Within the GAD framework we would expect that the experience of Cape Verde women would be shaped by these forces. We would also expect that their response would be a unique result of their history and social context.

This book also represents a fourth stage in the advance of ideas about women and development. This fourth stage is WCD, *Women, Culture, and Development*. WCD scholars criticize WID, WAD, and GAD for not taking culture into account. WCD advocates recommend that we put culture at the center of our assessment of women and their experience (Bhavnani et al., 2003). They define culture as the "as the lived experiences and material and emotional contexts that form the fabric of people's lives" (Bhavnani et al., 2003, p. 6). WCD advocates argue that women, oppression, resistance, and culture are all tied together and that in order to get a full picture of all of these forces, we must look at the everyday lives of individuals, especially those individuals on the periphery (Naples & Desai, 2002). WCD moves away from the singular economic focus of development to reveal the real complexities of third world communities. Dr. Bhavnani explains, "We need to attend to women because when you look at women, you see the complexities of people's lives. It is in women's lives that you see how the public and the private, or everyday life, are enmeshed—when you look at everyday life you are able to see agency" (Scott, 2008).

The GAD approach demands that we look at complex systems of power. Women and gender inequity are at the center but they are part of a dynamic, contradictory, and overlapping set of relationships which create oppression. Our globalized international economy of rich and poor, North and South, includes gender inequity as a central force as well.

WCD also reminds us that not only oppression but resistance is a central feature of our world. If we are to see (and nourish) that resistance, however, we must look beyond the strictly economic factors to also include the cultural expressions, the threads of every day life that make up our lives. Patricia Hill Collins (1990) has explored these complex systems of power that include economics and gender, politics and culture, oppression and resistance. Her theory of intersectionality provides a way of looking at the multiple layers of power, the multiple forms of resistance, and the interaction among them.

Complex Systems of Power

Collins (1990) writes that oppression is organized in a complex manner. First, she maintains there are several systems of inequality such as race ethnicity, gender, and social class. The social structures in which we live our lives are multi-faceted and contradictory. This means that every individual confronts and engages with others as a member of many different social categories. In the case of the Cape Verdean women of this study, she is a woman living in a structure of gender expectations and relationships, a citizen of Cape Verde, a resident of the Global South, a product of a long history of colonialism, slavery, and globalization and so forth.

Second, every individual experiences all of these systems simultaneously as we are all categorized in particular ways and relate to others because of our "categories" in different ways within each of these systems and their intersections. Expectations and behaviors regarding who we are in terms of race ethnicity or gender shape our lives and what is expected when we interact with others. Within the complex systems of inequality are webs of multiple relationships among people interacting within those systems.

Third, Collins asserts that people are not "pawns" within these systems randomly moved around a board. Rather they are actively responding to or resisting their place and the dominant sets of relationships. The complex systems of power relationships and structures are not static. They are dynamic and tensions result in pushing and pulling, creating change in those relationships and structures.

Collins asserts that our resistance is made further complex by occurring on three levels: "[1] the level of personal biography; [2] the group or community level of the cultural context created by race, class, and gender; [3] and the systemic level of social institutions" (1990, p. 114). By this she means that individuals are actively engaged in living their lives. They attempt to cope with, challenge or change the direction the system seems to be pushing them. As these individuals interact with one another coping, challenging, and changing, they create cultures of resistance and finally their actions and interactions ultimately transform the social context in which they find themselves. All three sites—biography, community, and system—are critical and the layered complexity as a whole is essential to our understanding.

Collins (1990) writes, "black feminist thought emphasizes all three levels as sites of domination and as potential sites of resistance" (p. 114). In this text, we have focused on all three of these as critical to understanding life in Cape Verde. On the first level of biography we have recorded the voices of Cape Verdeans as they described the economic and political context of their lives and their efforts to understand and use the policy tools to stop violence against women. We have listened to the women of Cape Verde talk about the problems they face and the ways they have coped with or resisted those difficulties.

At the cultural level, we have noted the importance of dance and language in disclosing life in Cape Verde and the manner in which women create and document their experience. And at the level of institution, we have provided a description of the political and economic character of globalization as well as the system of gender inequality that form the context of women's lives. When we put all of these together we conclude that culture is the key to understanding Cape Verde. As the WCD theorists maintain, an understanding of the political and economic context is essential but it is not enough to tell us about both the oppression of women in Cape Verde and of their resistance. Culture is key.

Culture Is Key

Colonial powers used a number of violent ways to control the people they dominated, such as war, imprisonment, enslavement, and exile. All of these are part of Cape Verde history. Today, neocolonial powers such as the World Bank and IMF have used other forms, such as SAPs and PRSPs that restrict social spending on schools and health care and increase poverty through low wages and restrictions of workers rights. Some scholars, however, argue that control in both historical and contemporary situations has been largely exercised through more subtle forms that control thinking and expression of ideas (Foucault, 1980). This control of the expression of ideas might be considered the control of culture. UNESCO (2002) defines culture in the following statement in their documents on cultural diversity: "culture should be regarded as the set of distinctive spiritual, material, intellectual and emotional features of society or a social group, and that it encompasses, in addition to art and literature, lifestyles, ways of living together, value systems, traditions and beliefs."

Helen Callaway (1987) argues that control of these features of human societies, control of culture, has historically and continues today to play a key role in the political dominance of powerful nations over others. She writes,

> [t]he case might be argued that imperial culture exercised its power not so much through physical coercion, which was relatively minimal though always a threat, but through its cognitive dimensions: its comprehensive symbolic order which constituted permissible thinking and action and prevented other worlds from emerging. (p. 87)

Amílcar Cabral (2008), an intellectual and political leader of the movement for national liberation of Portuguese colonies in the mid-twentieth century, also believed that culture was a key to social change (Chilcote, 1991; Chabal, 1983). In a speech given in 1970 at Syracuse University, Cabral outlined his theory of the value of culture arguing that culture was a powerful influence on human lives and that it can, therefore, act as both a vehicle for controlling people and for liberating them. Cabral (2008) was concerned that culture could be used by imperialists to hinder or prevent the development of forces to challenge domination when imperialist forces suppressed culture and thereby suppressed people.

Cabral (1970) believed that the way that colonial powers dominate is to eliminate the cultural life of the colonized people by preventing its expression or degrading it while simultaneously forcing colonized people to adopt the culture of the colonial power. He wrote,

> [t]he experience of colonial domination shows that, in the effort to perpetuate exploitation, the colonizer not only creates a system to repress the cultural life of the colonized people; he also provokes and develops the cultural alienation of a part of the population, either by so-called assimilation of indigenous people, or by creating a social gap between the indigenous elites and the popular masses.

In the case of Cape Verde, *batuku* was made illegal and stigmatized as shameful and primitive by the colonial powers. Creole was also degraded as a lesser form of expression not as sophisticated or proper as its European root, Portuguese. Cabral, like Collins (1990) also maintained that culture was not just a tool of the oppressor. Liberation movements could and must use the force of culture to resist domination:

> The value of culture as an element of resistance to foreign domination lies in the fact that culture is the vigorous manifestation on the ideological or idealist plane of the physical and historical reality of the society that is dominated or to be dominated. Culture is simultaneously the fruit of a people's history and a determinant of history, by the positive or negative influence which it exerts on the evolution of relationships between man and his environment, among men or groups of men within a society [and we would add women], as well as among different societies. Ignorance of this fact may explain the failure of several attempts at foreign domination—as well as the failure of some international liberation movements. (Cabral, 1970)

Our research shows that Cape Verdeans are heeding his advice, using the *batuku* dance and Creole proverbs as a "vigorous manifestation" of their experience and view of historical and contemporary oppression. Cabral says these are forms of resistance against foreign dominators. Today, women in Cape Verde are using them to challenge foreign dominators of the global economy. But they are also using them to defy systems of social class and gender dominance within their own nation.

Roads to Resistance

Our observations of women in Cape Verde have shown a difficult economic and social experience historically during slavery and colonialism and today as part of a global system of neocolonialism. The difficulties, however, have been met by resistance as people challenge the power structures on all levels. In particular, we have emphasized the middle level site of resistance which Collins identifies as: "the group or community level of the cultural context created by race, class, and gender."

Collins (1990) describes this site:

> The cultural context formed by those experiences and ideas that are shared with other members of a group or community which give

> meaning to individual biographies constitutes a second level at which domination is experienced and resisted. Each individual biography is rooted in several overlapping cultural contexts—for example, groups defined by race, social class, age, gender, religion, and sexual orientation. The cultural component contributes, among other things, the concepts used in thinking and acting, group validation of an individual's interpretation of concepts, the "thought models" used in the acquisition of knowledge, and standards used to evaluate individual thought and behavior. The most cohesive cultural contexts are those with identifiable histories, geographic locations, and social institutions. (p. 286)

The lives of the women in Cape Verde illustrate this cultural component. The words they use to conceptualize their lives and the language and dance they use to express their feelings about their lives and especially about the problems they face, the source of those problems and the need for change are all examples of the pieces of the cultural context Collins describes. Dancing and singing *batuku* gives meaning to women's lives. But dance and song also create and give meaning to cultural context. The women are preserving their own history, one that is quite different from history, as it has been constructed by colonial and contemporary global powers. Through *batuku* they are using words and physical movements that reflect the colonial history but not from the point of view of the colonialist. *Batuku* presents an example of the cultural context of resistance. Cape Verdean women challenge the colonial and neocolonial powers, as well as the power of men through dancing and singing the *batuku* and through expressions in Creole. Our study found that *batuku* is an arena of power struggle. Cape Verdeans express their African-ness despite attempts to abolish the dance or degrade it. *Batuku* is also a vehicle for expressing women's challenge to men and masculine hegemony by allowing the women to criticize men and express their sexuality.

Creole is similarly a tool for saving the Cape Verdean story from the perspective of oppressed people, retelling history, and expressing ideas that might not be acceptable to the dominant powers. The Creole language was forged by the forced intersection of the lives of Portuguese sailors and West Africans slaves. But today it is the language of the Cape Verdean people, one that is looked down on by those who celebrate their European ties and one that allows the Creole speakers to hide, express, and preserve their own ideas. Creole has preserved a memory of self and community identity. Creole has allowed women to express, cope with and resist the power structures of gender, and political and economic hierarchies in which they are the have-nots.

Women in Cape Verde are part of the international and continental drive to challenge governments to develop and implement policies to support women. They have been quite successful in regard to developing policies and ensuring that Cape Verde is formally adopting those policies. But these efforts are fairly low key. Along with the relatively quiet demands being made in these formal political arenas, the everyday acts of resistance in cultural forms, such as Creole and *batuku* are omnipresent and they are essential to building a developed revolutionary movement.

Finding Resources from Within

Firstly, resistance clearly accompanies all forms of domination. However, it is not always identifiable through organized movements; resistance inheres in the very gaps, fissures, and silences of hegemonic narratives. Resistance is encoded in the practices of remembering, and of writing. Agency is thus figured in the small day-to-day practices and struggles of third world women (Mohanty, 2003, p. 83).

A superficial reading of life in Cape Verde might lead us to anticipate that people would be struggling in the streets in light of the difficulties created by the global economic and political power structure. And we might wonder why they are not and attempt to explain the vacuum of resistance as a result of apathy or lack of leadership and organization. But perhaps we are not looking at the right things. Mohanty (2003) suggests that we "think through questions of resistance anchored in the daily lives of women" (p. 8). By this, she means that we need to look for the expression of resistance in the real women's lives rather than pasting on a notion of "revolutionary struggle" from other places and times.

Mohanty, like Cabral and Collins, suggests that an essential task in creating social change is "decolonizing and politicizing of knowledge by rethinking self and community through the practice of emancipatory education" (p. 10). This emancipatory education might take many forms and use many resources. Mohanty asserts that testimonials, life stories, and oral histories are significant modes of remembering and recording experience and struggles.

The transition from facing economic and political oppression to taking action to challenge the power structure in full-blown revolutionary activity is a complex one that must gather forces from many sources and it must gather from the sources that are available. Collins (1990) writes that a first step may come through cultural resources. In the case of African American women, for example, it might come from blues music, a deep rooted historical resource in the African American community. She writes,

> Self-defining in these ways allows Black women to move from victim to freedom to action. Blues music empowers women to act by using their individual voices to sing about issues experienced collectively. (Collins 1990, p. 100)

Self-definition through music, literature, and community provides Black American women with "the spirit of independence," skills to become "self-reliant," and the support and encouragement required to challenge traditional stereotypes about black women (p. 109). By creating their own definitions, black American women place "the power to save the self within the self," rather than looking outside for rescue (p. 112). Similarly, in Cape Verde, the women draw on the historical cultural resources of their community, of Creole and *batuku* which allows them to define themselves and to position themselves in opposition to the forces that oppress them.

The process of self-definition is also one of consciousness-raising. This creates the locus for social change, that is, to be valued as black women by their own definition, rather than to live the struggle "of two lives, one for them [the concept of the Black woman as defined by the dominant culture] and one for ourselves" (Gwaltney quoted in Collins, 1990, p. 94). In Cape Verde we could interpret the speaking and dancing as a way of educating oneself and others about what is happening and why. In addition, these forms of expression, Creole, and *batuku* create vehicles "of oppositional agency" (Mohanty, 2003, p. 52).

Reconceptualizing Resistance

The findings of this research demand a reconceptualization of the concept of resistance that incorporates the ideas of Collins, Cabral, and Mohanty. Resistance is the action of an individual or individuals opposing something that they disapprove of. When we think of resistance against globalization, we might think of international protests at the G20 meetings, of Bolivian peasants joining together to stop the privatization of water, French farmers revolting against trade policies harming small farms, and the Mayan people in Mexico organizing to defend the rights of indigenous people (Cavanagh & Mander, 2004). In addition, women in Africa have been showing resistance in a collective manner for centuries to fight for other issues such as land rights, employment, health care, and rights of workers.

Women, however, are often excluded from formal power structures such as elected office or leadership in international organizations and NGOs. Women may, therefore, choose to or be forced to employ other forms of political action, political action that takes the form of cultural creation and expression. Dance, proverbs, and sayings are alternative forms of resistance. Through these, women are able to expose and discuss sensitive public and private political issues such as infidelity, oppression, and corruption that would not normally be allowed in everyday direct speech. Through these activities women acknowledge their difficulties, identify their causes, and construct different possibilities in their world.

Important Lessons

Observing this expanded array of political resistance is important because it opens our eyes to the significance of women's everyday activities we might not immediately recognize as such. While many people might be interested in *batuku* and Creole as entertaining cultural expressions, we might not see the ways in which these forms of expression are part of a political exchange among the participants contemplating the unjust distribution of power.

Taking a broader view of resistance to include these everyday acts of rebellion also calls into question the limitations of our conceptualization of political resistance as only fitting into certain preconceived categories such as marches, demonstrations, strikes, and elections. Nikki Giovanni argues, "people are seldom powerless, no matter how severe the limitations in their lives." Observing the women of Cape Verde forces us to redefine resistance and politics to include relationships and actions outside of the formal arenas of power (Kutsche & Ness, 1981).

Finally, broadening our lens to rethink what dancing and Creole sayings might express also allows us to see more clearly the ordinary lives of women and the extraordinary resources they can call upon to face the challenges of globalization. Taking a broader look allows us to see paths to social change we might otherwise overlook.

According to Collins (1990) both the changed consciousness of individuals and the social transformation of political and economic institutions constitute essential ingredients for social change. New knowledge is important for both dimensions to change because knowledge is a vitally important part of the social relations of domination and resistance and new knowledge is generated by new forms of consciousness. One is feminist consciousness, which is the "reflection on women's experience drawing attention to pervasive patterns of subordination, limitation, and confinement. Within this consciousness, alternative visions for living can develop. Recognizing alternative worldviews, rather than one hegemonic view, is key to understanding women's experience" (Keohane et al., 1982, p. x).

The women of Cape Verde are creating or at least preserving knowledge about themselves and their lives through a broad range of expressions in Creole sayings and *batuku*. They are raising consciousness among themselves by sharing their experiences and their opinions. On the basis of Keohane and Collin's arguments, Cape Verdean women are developing a consciousness that will allow them to (or demand that they) consider alternatives. And that reconsideration is essential to creating political demands and political movements.

They are not yet transforming social institutions through these activities. But they are creating paths to such a transformation. Moving from acknowledging oppression and identifying the source of the problem to explicit political activity that seeks to transform social and political institutions is a huge leap. Identifying the bridges between these is critical. *Batuku* and Creole among women in Cape Verde may be just such a link.

In order to make changes in women's lives and their communities, a first step is to become aware of a different world view, one that challenges the dominant view which keeps women in one place. Alternative forms of expression such as dancing and alternative use of language preserve and express alternative ways of thinking and acting.

From Resistance to Transformation

If we are to stop globalization from creating misery in our lives and instead find ways to make it improve life for everyone on the planet, we will need to move beyond resistance. We will need to move to transformation, which will take organized political movements. But in order to develop political movements, we need to find bridges to those transformative organizations and activities. Everyday life will provide the tools to build those bridges. Our scholarship and theories need to uncover those tools.

Nancy Hartsock writes,

> we need a theory of power that recognizes that our practical daily activity contains an understanding of the world—subjugated perhaps, but present.... [In addition,] our understanding of power needs to recognize the difficulty of creating alternatives. The ruling class, race and gender actively structure the material-social relations in which all the parties are forced to participate; their vision, therefore, cannot be dismissed as simply false or misguided. In consequence, the oppressed groups must struggle for their own understandings which will represent achievements requiring both theorizing and the education which grows from political struggle. (Hartsock 1990, pp. 158–172)

Alternatives are difficult to find because they must break through the hegemonic discourse which make the everyday activities of oppressed people appear at best mundane and at worst, weak and insignificant. In Cape Verde we propose that alternatives are coming from the daily activities of staying alive, speaking of our lives, and expressing our feelings through cultural vehicles of music and dance.

Reflections on Chapter 9 Research Methods

This research is a qualitative study. One of the most important distinctions between qualitative and quantitative work is the relationship of theory to data in the two types (Creswell, 2003). Quantitative researchers begin with theory and work deductively to gather data and test hypothesis that are derived from the theories. In a report of quantitative research, theory is discussed near the beginning. Qualitative researchers work inductively, starting with the empirical evidence and working toward a theoretical conclusion. Qualitative researchers enter with a set of goals and questions that guide the research but must be open to what comes up along the way (Lareau, 2003). In fact, what comes up along the way is the core of a qualitative study. Theory then helps us to sort out the meaning and significance of what we have found along the way. This chapter on theory, therefore, appears at the end of our book.

This chapter attempts to summarize and make sense of what we have learned in our study. We begin by noting what our initial research questions were and summarizing what we learned and have presented in the previous chapters. We then identify the core discoveries of our investigation: we discovered that we had begun the research with certain assumptions about what oppression and resistance are and those assumptions were not entirely supported by the empirical findings. The data told us that our assumptions were limited and that we would need to rethink some of these concepts in order to make sense of the data. Thinking and rethinking concepts and trying to make sense of social issues and information are called theorizing. This chapter uses theory to tie all of the previous chapters together by trying to make sense of what we have observed.

If we only report information that describes events and social relationships, we have fallen short of scholarship (May, 2001). Tim May writes, "The idea of theory, or the ability to explain and understand the findings of research within a conceptual framework that makes "sense" of the data, is the mark of a mature discipline whose aim is the systematic study of particular phenomena" (p. 29). Scholars, of course, must describe the phenomenon they study. But they must go beyond that to reflecting on the information and offering some sense of what this all means. Qualitative research is especially interested in providing in-depth description of all of the details of the subject of their study. But they also must take that information and explore its abstract meaning in order to make a contribution to our theoretical understanding.

G.A. Cohen (1984) explains that theory is not so mysterious as we sometimes might believe. Theory is only making connections between the real world and the ways we think about it (May, 2001). Cohen (1984) writes, "Theory aims at the production of thought that is in accord with reality. Practice aims at the production of realities which accord with thought. Therefore common to theory and practice is an aspiration to establish congruity between thought and reality" (p. 339).

In this chapter we use Collin's theory of intersectionality to explore and explain the connections among the abstract concepts that emerged as most significant in our study. We asked what are globalization, oppression, resistance, gender, and culture? And how are they connected with one another? Our goal was to understand the connections between the practice (the experience of Cape Verdeans in a globalized world as they expressed it to us) and the ways we think about oppression, resistance, gender and culture.

This chapter has also left us with some ideas about what the practical application of our theoretical understanding might be. As we mentioned in chapter 3, this study fits within a critical framework. That means that we must go beyond description and understanding to present suggestions for how the research can help to transform our unjust world. "Critical social research assumes that the world is changed though reflective practical activity and is thus not content to simply identify the nature of oppressive structures but to point to ways in which they can be combated through praxis" (Harvey, 1990, p. 32). We suggest that taking a broader view of these issues, based on our reading of the empirical data, and identifying the connections among globalization, oppression, resistance, gender, and culture are not only important to further our thinking on the issues. Our understanding of the concepts and their connections are important to our building social movements that are rooted in the social context of real people's lives and reflect not only the political and economic features of globalization but also its gendered character and the role of culture in promoting social change.

Appendix A

Questionnaire: Economics and Politics

1. Since 1991, (the change to multiparty elections), how have things changed for rich people in Cape Verde?
2. Since 1991, how have things changed for poor people in Cape Verde?
3. How has the relationship between rich and poor changed since 1991?
4. Can you think of an example that shows how the gap between rich and poor has grown larger since 1991?
5. Do you think poverty is a bigger problem than it was in 1991?
6. What should the government do to help solve the poverty problem?

Appendix B

Questionnaire: *Batuku* Dance

1. Why is the *batuku* considered the national dance of Cape Verde?
2. What is significant about the *batuku*?
3. Do you remember learning it as a child? How was its form/movements and its history explained to you?
4. The *batuku* was banned during Colonization though people still continued to dance it in private. Why did they do so?
5. Are there other issues about the *batuku* you want to mention?

Appendix C

Questionnaire: Creole Language

1. Some people say that if a person does not know *Creole*, they do not know Cape Verde. Do you agree with this statement? Explain why you do or do not agree with this statement.
2. If someone only speaks Portuguese in Cape Verde, what exactly might they miss? Try to think of a specific example of something a person would miss if they only spoke Portuguese.
3. Do you think it is important for a person who comes to Cape Verde to learn *Creole* to get along? Why?

Appendix D

Interview Questions: Cape Verdean Women

1. Background information: age, where born, where family lives, formal schooling, job, children, who raises your children, who were you raised by?
2. Women's Work: How do you make a living? Is it sufficient? Can you get by? Are you primarily responsible for feeding your family? How do you feel about your work?
3. Relationships with Men: Do the men in your life live up to your expectations? Why or why not? Do they provide you economically? Emotionally? How do you feel when your husband/mate/boyfriend begins a relationship with another woman? Do women also have other men partners—more than one partner?
4. Having Children: Do you have children? Do your children provide emotional and financial support for you or are they a burden? If your husband does not live with you, do you receive child support? Is having many children a good idea or a bad idea? What is your opinion/attitude toward family planning/birth control/condoms?
5. Family and Friends as support: Do your family and friends help you—economically and emotionally? How? Do you depend on them? How? What is your relationship with your Godmother? Has she helped you? How? Do you have a Godchild? What is your relationship with him/her? Have you helped him/her? How?

Appendix E

Questionnaire: Women's Rights Conference

1. I learned something today that I can apply to my life. Write down a number between 1 and 5. (1 = very little 5 = a lot)
2. What did you learn today that you could apply to your life?
3. What was the best part of the conference?
4. What would you change about the conference?
5. In your opinion, what are 2 problems women experience?
6. What are 2 recommendations or ideas for solutions?
7. What are the most important problems/barriers you face in trying to create positive change?
8. What external help or factors would benefit your organization in addressing/solving problems?
9. Have you attended previous women's conferences? If yes, how are conferences useful for your work on women's issues?
10. What is especially important to include in a women's conference to make it useful and practical?

Bibliography

Abugre, C., 2001. *Still sapping the poor: A critique of IMF poverty reduction strategies* [Online]. Available at: www.wdm.org.uk [accessed September 2, 2008].

ACHPR, 2008. Protocol to the African Charter on Human and Peoples' Rights on the Rights of Women in Africa [Online]. Available at: www.achpr.org [accessed November 25, 2008].

Acosta-Bélen, E. & Bose, C.E., 1995. Colonialism, structural subordination, and empowerment. In C. Bose & E. Acosta-Belén, eds. *Women in the Latin America Development Process.* Philadelphia: Temple University Press. 15–36.

Adams, M., 2006. Regional women's activism: African women's networks and the African Union. In M.M. Ferree & A. Tripp, eds. *Global feminism: Transnational women's activism, organizing and human rights.* New York: New York Press. 187–218.

Adande, J.O. & Kuye, P.O., 1986. Nigeria: Family law project. In M. Schuler, ed. *Empowerment and the law: Strategies of third world women.* Washington, DC: OEF International. 335–345.

Africa Research Bulletin, 2007. No longer poor? *Africa Research Bulletin*, 44 (11), 17615–17616.

———, 2006. Emigrant remittances. *Africa Research Bulletin*, 43 (10), 17140.

Africa—South of the Sahara, 26th edition, 1997. London: Europa Publications.

Afrol, 2006. *Tourism boom fuels Cape Verde economy* [Online]. Available at: www.afrol.com/articles/24742 [accessed November 9, 2008].

———, 2003a. *Privatized Cape Verdean utilities increase tariffs* [Online]. Available at: www.afrol.com [accessed September 4, 2008].

———, 2003b. *Protests against Cape Verdean price hikes* [Online]. Available at: www.afrol.com [accessed September 4, 2008].

———, 2002. *WFP famine aid appeal for Cape Verde* [Online]. Available at: www.afrol.com/News2002/cav019_wfp_appeal.htm [accessed November 10, 2008].

Agar, M., 1980. *The professional stranger: An informal introduction to ethnography* New York: Academic Press.

AGIPNEWS, 2008. *Cape Verde becomes the WTO's 153rd member* [Online]. Available at: www.ag-ip-news.com [accessed August 1, 2008].

Ahmed, G., 1994. *Female-labor migration and emancipation of women in Southern Africa: Fact or fiction?* Presented at the annual meetings of the International Sociological Association. Bielefeld, Germany. July 18–23.

Alasuustara, P., Bickman, L., & Brennan, J., 2009. Introduction: Changing research in changing social conditions. In P. Alasuurara, L. Bickman, & J. Brennan, eds. *Sage handbook of social research methods.* Thousand Oaks: Sage.

Albrecht, C., 2001. Sex ratio and family structure in the non-metropolitan United States. *Sociological Inquiry,* 71 (1), 67–84.

Allen, E., 2005. Intrapersonal, interpersonal, and contextual factors in engaging in and responding to extramarital involvement. *Clinical Psychology: Science and Practice*, 12, 101–130.

Almeida, R., 1997a. *Chronological references: Cabo Verde/Cape Verdean American* [Online]. Available at: www.umassd.edu/specialprograms/caboverde [accessed July 1, 2006].

———, 1997b. *The church and the people of Cape Verde* [Online]. Available at: www.umassd.edu/specialprograms/caboverde [accessed July 1, 2006].

———, 1996. *Fishing for a living* [Online]. Available at: www.umassd.edu/specialprograms/caboverde [accessed June 15, 2006].

———, undated. *Cape Verdeans in the whaling industry*, undated [Online]. Available at: www.umassd.edu/specialprograms/caboverde [accessed June 15, 2006].

Alsop, Z., 2008. *Kenya farmers reap profits sown by joined hand* [Online]. Available at: www.womensenews.org [accessed November 23, 2008].

Ames, B., Renno, L., & Rodrigues, F., 2003. *Democracy, market reform and social peace in Cape Verde.* Cape Town: Afrobarometer.

Andrade, L.L.D., 2000. Negotiating from the inside. *Journal of Contemporary Ethnography*, 29 (3), 268–290.

Andre, R., 1979. *Multi-cultural research: Developing a participative methodology for cross-cultural research.* Presented at the Annual Convention of the International Council of Psychologists, Princeton, NJ.

Angen-Urdinola, D. & Wodon, Q., 2008. *Assessing the targeting performance of social programs: Cape Verde.* Washington, DC: World Bank.

AnimalInfo, 2005. *AnimalInfo—Cape Verde* [Online]. Available at: www.animalinfo.org/country [accessed June 15, 2006].

Argus, C., 2004. Cancel Africa's debt, UN Says. *Cape Argus,* September 30, 15.

Ashkenazi, M., 1997. Informant networks and their anthropologists. *Human Organization,* 56 (4), 471–477.

Aulette, J., Wittner, J., & Blakely K., 2009. *Gendered worlds.* New York: Oxford University Press.

Ault, A. & Sandberg, E., 2000. Our policies, their consequences: Zambian women's lives under structural adjustment. In L. Richardson, V. Taylor, & N. Whittier, eds. *Feminist frontiers V.* New York: McGraw-Hill. 503–506.

Baca Zinn, M., 2001. Insider field research in minority communities. In R. Emerson, ed. *Contemporary field research*, 2nd edition. Prospect Heights: Waveland. 159–166.

———, 1979. Field research in minority communities: Ethical, methodological and political observations by an insider. *Social Problems,* 27, 209–19.

Baker, B., 2006. Cape Verde: The most democratic nation in Africa? *Journal of Modern African Studies*, 44 (4), 493–511.

Barata, P., McNally, M.J., Sales, I., & Stewart, D., 2005. Portuguese-speaking women voice their opinions: Using their words to teach about wife abuse. *Women's Health Issues*, 15, 134–144.

BBC News, 2008. Country profile: Cape Verde. June 4 [Online] Available at: news.bbc.co.uk/ [accessed October 30, 2008].

———, 2006. Q&A: Cape Verde presidential election. February 10 [Online]. Available at: news.bbc.co.uk/ [accessed November 10, 2008].

Beoku-Betts, J., 1994. When black is not enough: Doing field research among Gullah women. *NWSA Journal,* 6, 413–433.

Bhavnani, K.K., Foran, J., & Kurim, P., 2003. An introduction to WCD. In K. Bhavnani, J. Foran, & P. Kurim, eds. *Feminist futures.* London: Zed.

Bickerton, D., 1981. *Roots of language.* Ann Arbor: Karoma.

Bigman, L., 1994. *History and hunger in West Africa: Food production and entitlement in Guinea-Bissau and Cape Verde.* Westport: Greenwood.

———, 1993. *History of hunger in West Africa.* Westport: Greenwood.

Blauner, R. & Wellman, D., 1973. Toward the decolonization of social research. In J.A. Ladner, ed. *The death of white sociology.* New York: Vintage Books. 310–330.

Boonzaier, F., 2005. Women abuse in South Africa: A brief contextual analysis. *Feminism & Psychology*, 15, 99–103.

Boserup, E., 1970. *Woman's role in economic development.* London: Earthscan.

Boyatzis, R., 1998. *Transforming qualitative information.* Thousand Oaks: Sage.

Bozalek, V., 2004. Subjugated knowledges and the South African social work curriculum: Students' accounts of the impacts of racism, gender and generation on their family practices. *Global Social Work Congress,* Adelaide, Australia.

Bradley, D., 2002. Language attitudes: The key factor in language maintenance. In D. Bradley & M. Bradley, eds. *Language endangerment and language maintenance.* London: Routledge Curzon. 1–10.

Braithwaite, R., Cockwill, S., O'Neill, M., & Rebane, D., 2007. Insider participation action research in disadvantaged post- industrial areas: The experiences of community members as they become community based action researchers. *Action Research,* 5 (1), 61–74.

Bransford, C., 2006. The use of critical ethnography in managed mental health care. *Journal of Sociology and Social Welfare*, 33 (4), 173–191.

Braun, F., 2001. The communication of gender in Turkish. In M. Hellinger & H. Bussmann, eds. *Gender across languages: The linguistic representation of women and men.* Amsterdam: John Benjamins. 283–310.

Brenzinger, M., Heine, B., & Sommer, G., 1991. Language death in Africa. In R. Robins & E. Uhlenbeck, eds. *Endangered languages.* Oxford: Berg. 19–44.

Brettell, C.B. & Sargent, C.F., 1993. Gender, ritual, and religion. In C.B. Brettell & C.F. Sargent, eds. *Gender in cross-cultural perspective*. Englewood Cliffs: Prentice-Hall. 246–250 and 332–336.

Brydon, L. & Chant, S., 1989. *Women in the third world: Gender issues in rural and urban Areas*. New Brunswick: Rutgers University.

Bryson, B., 1990. *The mother tongue—English & how it got that way*. New York: HarperCollins.

Bujra, J., 1986. Urging women to redouble their efforts: Class, gender, and capitalist transformation in Africa. In C. Robertson & I. Berger, eds. *Women and class in Africa*. New York: Africana. 117–140.

———, 1979. Introduction: Female solidarity and the sexual division of labor. In P. Caplan & J. Bujra, eds. *Women united, women divided: Comparative studies of ten contemporary cultures*. Bloomington: Indiana University. 13–45.

Cabral, Amílcar, 2008. *Unity and struggle*, 2nd edition. Johannesburg: UNISA Press.

———, 1970. *History is a weapon: National liberation and culture*. Speech at Syracuse University: New York. February 20.

Cagatay, N., 2003. Gender budgets and beyond: Feminist fiscal policy in the context of globalisation. *Gender and Development*, 11 (1), 15–24.

Callaghy, T., 1993. Vision and politics in the transformation of the global political economy: Lessons from the second and third worlds. In R.O. Slater, B.M. Schutz, & S.R. Door, eds. *Global transformation and the third world*. Boulder: Lynne Rienner. 161–247.

Callaway, H., 1987. *Gender culture and empire: European women in colonial Nigeria*. Urbana: University of Illinois Press.

Campaign for Labor Rights, 2003. *Reforming the IMF and the World Bank to Millennium Development Goals—statement by global unions to the 2003 annual meeting of the IMF and World Bank* [Online]. Available at: www.campaignforlaborrights.org [accessed November 1, 2005].

Campbell, C., 1992. Learning to kill? Masculinity, the family and violence in Natal. *Journal of Southern African Studies*, 18 (3), 614–628.

Cannon, L.W., Higginbotham, E., & Leung, M.L.A., 1988. Race and class bias in qualitative research on women. *Gender and Society*, 2, 449–462.

Cape Verde Agricultural Census. 1988 [Online]. Available at: www.fao.org [accessed November 1, 2005].

Cape Verdean Foods, undated. [Online]. Available at: www.umassd.edu [accessed November 1, 2005].

Carling, J., 2004a. Emigration, return and development in Cape Verde: The impact of closing borders. *Population, Space and Place*, 10 (2), 113–132.

———, 2004b. Figuring out the Cape Verde diaspora. *Cimboa*, 4 (2), 3–9.

———, 2002a. Cape Verde: Towards the end of emigration? [Online]. Available at: www.migrationinformation.org [accessed November 1, 2005].

———, 2002b. Migration in the age of involuntary immobility: Theoretical reflections and Cape Verdean experiences. *Journal of Ethnic and Migration Studies*, 28 (1), 5–43.

———, 2001. *Aspiration and ability in international migration: Cape Verdean experiences of mobility and immobility.* Dissertation. University of Oslo.

Carter K. & Aulette, J., 2009. Creole in Cape Verde: Language, identity, and power, *Ethnography*, 10 (2), June.

CATO Institute, 2008. Benefits of globalization. Washington, DC: Center for Trade Policy Studies [Online]. Available at: www.freetrade.org/issues/globalization.html [accessed October 30, 2008].

Cavanagh, J. & Mander, J., eds., 2004. *Alternatives to economic globalization.* San Francisco: Berrett-Koehler.

Center for Disease Control, 2008. U.S. Public Health Services syphilis study at Tuskegee [Online]. Available at: www.cdc.gov/tuskegee/timeline.htm [accessed November 5, 2008].

Chabal, P., 1996. The transition to multi-party politics in Lusophone Africa: Problems and prospects. *Lusotopie*, 57–69.

———, 1983. *Amilcar Cabral: Revolutionary leadership and the people's war.* London: Cambridge University Press.

Chant, S., 2001. Men in crisis? Reflections on masculinities, work and family in North-West Costa Rica. In C. Jackson, ed. *Men at work: Labour, masculinities, and development.* London: Frank Cass.

Charmaz, K., 2006. *Constructing grounded theory: A practical guide through qualitative analysis.* Thousand Oaks: Pine Forge Press.

Cheek, J., 2004. At the margins? *Qualitative Health Research*, 14, 1140–1150.

Chelala, C., 2000. Concern grows about adolescent pregnancy in Cape Verde. *Lancet,* 355, 128–138.

Chen, M., 1995. The feminization of poverty. In N. Heyzer, ed. *A commitment to the world's women: Perspectives on development for Beijing and beyond.* New York: UNIFEM.

Chilcote, R., 1991. *Amílcar Cabral's revolutionary theory and practice.* London: Lynne Reinner.

Choi, S. & Ting, K.F., 2008. Wife-beating in South Africa: An imbalance theory of resources and power. *Journal of Interpersonal Violence*, 23, 834–852.

CIA World Fact Book, 2008 [Online]. Available at: www.cia.gov [accessed November 10, 2008].

———, 2006. [Online]. Available at: www.cia.gov [accessed May 10, 2007].

Civil G8, 2006. *World on track to halve poverty rate by 2015 but Africa lags well behind: A report* [Online]. Available at: www.en.civilg8.ru [accessed November 1, 2005].

Clark, V., Eschholz, P., & Rosa, A., 1998. *Language—readings in language and culture,* 6th edition. Boston: Bedford/St. Martin's Press.

Coates, J. ed., 1998. *Language and gender: A reader.* Oxford: Blackwell.

Cohen, G.A., 1984. *Karl Marx's theory of history: A defence.* Oxford: Oxford University Press.

Coleman, R. & Daniels, J., 2000. Mediating Ebonics. *Journal of Black Studies,* 31, 74–95.

Collier, J. & Collier, M., 1986. *Visual anthropology: Photography as research method.* Albuquerque: University of New Mexico Press.

Collins, P.H., 2000. *Black feminist thought: Knowledge, consciousness, and the politics of empowerment*. New York: Routledge.

———, 1990. *Black feminist thought: Knowledge, consciousness, and the politics of empowerment*. Boston: Unwin Hyman.

Commission on Human Rights, 2003. *Integration of the human rights of women and the gender perspective: Violence against women* [Online]. Available at: www.hri.ca/fortherecord2003 [accessed July 30, 2008].

Condon, J., 1986. So near the United States. In J.M. Valdes, ed. *Culture bound—bridging the cultural gap in language teaching*. Cambridge: Cambridge University Press. 85–93.

Copelon, R., 1994. Intimate terror: Understanding domestic violence as torture. In R. Cook, ed. *Human rights of women: National and international perspectives*. Philadelphia: University of Pennsylvania. 116–152.

Corbin, J. & Strauss, A., 2008. *Basics of qualitative research*, 3rd edition. Thousand Oaks: Sage.

CountryWatch, 2007. Cape Verde: Country review [Online]. Available at: www.CountryWatch.com [accessed November 1, 2008].

Creswell, J.W., 2003. Research design: Qualitative, quantitative, and mixed methods approaches, 2nd edition. Thousand Oakes: Sage.

Curtin, P., 1969. *The Atlantic slave trade: A census*. Madison: University of Wisconsin Press.

da luz Goncalves, M., 2001. Kriolu is more than a language. *Footsteps*, 3 (4), 15–18.

Davidson, B., 1989. *The fortunate isles: A study in African transformation*. Trenton, NJ: Africa World Press.

DeParle, J., 2007. In a world on the move, a tiny land strains to cope. *New York Times*, June 24.

DeVault, M., 1996. Talking back to sociology: Distinctive contributions of feminist methodology. *Annual Review of Sociology*, 22, 29–52.

———, 1995. Ethnicity and expertise: Racial-ethnic knowledge in sociological research. *Gender and Society*, 9, 612–631.

Dixon, R., 2004. *Cancel Iraqi debt? What about Africa?* [Online]. Available at: www.why-war.com [accessed November 1, 2005].

Drew, A., 1995. Female consciousness and feminism in Africa. *Theory and Society*, 24 (1), 1–33.

Dunnett, S., Dubin, F., & Lezburg, A., 1986. English language teaching from an intercultural perspective. In J.M. Valdes, ed. *Culture bound—bridging the cultural gap in language teaching*. Cambridge: Cambridge University Press. 148–161.

Ehrenreich, B. & Hochschild, A.R., eds., 2002. *Global woman: Nannies, maids, and sex workers in the new economy*. New York: Henry Holt.

Eisenstein, Z., 1998. *Global obscenities: Patriarchy, capitalism, and the lure of cyberfantasy*. New York: NYU Press.

Elson, D., 1992. Male bias in structural adjustment. In H. Afshar & C. Dennis, eds. *Women and adjustment policies in the third world*. New York: St. Martin's Press. 46–68.

Emasu, A., 2002. *Get the land, the rest will follow* [Online]. Available at: www.africawoman.net [accessed November 25, 2008].

Emerson, R., 2001. *Contemporary field research,* 2nd edition. Prospect Heights: Waveland.

Encontro national sobre a problematica da mulher Cabo-Verdeana. 1986. Praia.

Escure, G., 2001. Belizean Creole: Gender, creole, and the role of women in language change. In M. Hellinger & H. Bussmann, eds. *Gender across languages: The linguistic representation of women and men.* Amsterdam: John Benjamins. 53–84.

Europa World Year Book, 2003. Volume 1, London: Europa Publications.

Fage, J.D. & Olivier, R., 1978. *The Cambridge history of Africa.* Cambridge: Cambridge University Press.

Fairclough, N., 1992. *Discourse and social change.* London: Polity.

Fairclough, N. & Wodak, R., 1997. Critical discourse analysis. In T. Van Dijk, ed. *Discourse as Social Interaction.* London: Sage. 258–284.

Fairclough, N., Graham, P., Lemke, J., & Wodak, R., 2004. Introduction. *Critical Discourse Studies* 1 (1), 1–7.

Fajana, A. & Anjorin, A.O., 1979. *From colony to sovereign state: An introduction to the history of West Africa since 1900.* London: Nelson.

Ferguson, C.A., 1959. Diglossia. *Word,* 15, 325–40.

Finan, T. & Henderson, H., 1988. The logic of Cape Verdean female-headed households: social response to economic scarcity. *Urban Anthropology,* 17 (1), 87–103.

Fiske, J., 1994. *Media matters: Everyday culture and political change.* Minneapolis: University of Minnesota Press.

Flick, U., 2006. *An introduction to qualitative research*, 3rd edition. Thousand Oaks: Sage.

Foucault, M., 1980. *Power/Knowledge: Selected interviews and other writings, 1972–1977.* New York: Pantheon.

Fryer, P., 2000. *Rhythms of resistance: African musical heritage in Brazil.* Hanover: Wesleyan University Press.

Gonçalves, M., undated. *Cape Verdean Kriolu in the United States* [Online]. Available at: www.umassd.edu/SpecialPrograms/CaboVerde [accessed October 18, 2007] .

González, L. S., 1999. Reclaiming salsa. *Cultural Studies*, 13 (2), 237–250.

Green, D., 2000. *Gender violence in Africa—African women's responses.* New York: St. Martin's Press.

Greenspan, A., 2000. *Globalization. Speech to federal reserve board in Mexico City* [Online] Available at: http://www.federalreserve.gov/BoardDocs/Speeches [accessed October 30, 2008].

Halim, A.M.A., 1994. Tools of suppression. In A. Busia, ed. *Gender violence and women's human rights in Africa.* Center for Women's Global Leadership. Highland Park: Plowshares. 21–29.

Hartsock, N., 1990. Foucault on power: A theory for women? In L. Nicholson, ed. *Feminism/Postmodernism.* New York: Routledge. 158–172.

Harvey, L., 1990. *Critical social research*. LONDON: Unwin Hyman.
Heath, D., 1994. The politics of appropriateness and appropriation: Recontextualizing women's dance in urban Senegal. *American Ethnologist*, 21 (1), 88–103.
Heise, L. & Garcia-Moreno, C., 2002. Violence by intimate partners. In E.G. Krug, L.L. Dahlberg, J.A. Mercy, A.B. Zwi, & R. Lozano, eds. *World report on violence and health*. Geneva: World Health Organisation. 89–122.
Held, D. & McGrew, A., 2000. *The global transformation reader: An introduction to the globalization debate*. Cambridge: Polity Press.
Herbison-Evans, D., 1988. History of Latin-American dancing. *Australian Dance Review*, 39, 28–29.
Hewer, P., 2000–2001. Review, A.K. Awedoba, An introduction to Kasena society and culture through their proverbs. *The Journal of West African Languages*, 28, 117–120.
Holloway, I., 1997. *Basic concepts for qualitative research*. London: Blackwell.
Holmes, J., 1998. Women's talk: The question of sociolinguistic universals. In J. Coates, ed. *Language and gender: A reader*. London: Blackwell. 461–483.
Human Rights Watch. 1995. *Violence against women in South Africa*. New York: Human Rights Watch.
Humphrey, C., 2007. Insider outsider. *Action Research*, 5 (1), 11–26.
Hurley-Glowa, S. 2006. *Funana* with a drum machine beat. *The International Journal of Africana Studies*, 12, 80–92.
———, 2001. Dance to the music. *Footsteps* 3 (4), 24–28.
International Monetary Fund, 2008. *IMF members' quotas and voting power, and IMF Board of Governors* [Online]. Available at: www.imf.org [accessed October 30, 2008].
———, 2007. *The IMF at a glance* [Online]. Available at: www.imf.org [accessed November 1, 2008].
———, 2005. *Cape Verde: Poverty reduction strategy paper—joint staff advisory note*. Washington, DC: IMF.
IRIN news, 2004. *Cape Verde: Morocco sends planes to spray locusts in Cape Verde islands*. UN Office for the Coordination of Humanitarian Affairs [Online]. Available at: www.IRINnews.org [accessed May 10, 2005].
Irwin, A. & Wilson, C., 1998. *Cape Verde Islands*. Bucks: Bradt Travel Guides.
Ishemo, S., 1995. Cultural responses to forced labour and commodity production in Portuguese African colonies. *Social Identities*, 1 (1), 95–110.
Jensen, J., 2001. I'd rather be dancing: Wisconsin women moving on. *Frontiers: A Journal of Women Studies*, 22 (1), 1–20.
Jere-Mwiindilila, P., 1994. The effects of structural adjustment on women in Zambia. *Reformed World* [Online], 44 (2), Available at: www.warc.ch/pc/rw942/02 [accessed August 19, 2007].
Johnson, B., 2004. [Discussion] (Personal communication, February 5, 2004).
Johnstone, B., 2002. *Discourse analysis*. London: Blackwell.
Jones, P., 2004. Critical discourse analysis. *Historical Materialism* 12 (1), 97–125.
Kapoor, S., 2006. *Exposing the myth and plugging the leaks*. Social Watch Report [Online]. Available at: www.socialwatch.org [accessed May 10, 2007].

Keohane, N.O., Rosaldo, M.Z., & Gelpi, B.C., 1982. *Feminist theory: A critique of ideology*. Chicago: University of Chicago Press.

Kerr, J. & Sweetman, C., eds., 2003. *Women reinventing globalization*. London: Oxfam Press.

Krieger, S., 1985. Beyond subjectivity and the use of the self. *Qualitative Sociology*, 8, 309–324.

Kutsche, P. & V. Ness, 1981. *Canones-values, crisis, and survival in a Northern Mexico village*. Salem: Sheffield.

Languages of Cape Verde Islands, undated. Ethnologue report for Cape Verde Islands [Online]. Available at: www.ethnologue.com [accessed May 5, 2008].

Ladner, J., 1973. *The death of white sociology*. New York: Vintage Books.

Langworthy, M. & Finan, T., 1997. *Waiting for rain: Agricultural and ecological imbalance in Cape Verde*. Boulder: Lynne Reinner.

Lareau, A., 2003. *Unequal childhoods: Class, race, and family life*. Berkeley: University of California Press.

Levy, C., 1996. The process of institutionalizing gender policy. *Policy planning guide to gender analysis frameworks*. UK: Oxfam.

Lobban, R., 1998. *Cape Verde: Crioulo colony to independent nation*. Boulder: Westview Press.

———, 1996. *Jews in Cape Verde and on the Guinea Coast* [Online]. Available at: www.saudades.org [accessed November 1, 2005].

Lucy, J., 1997. Linguistic relativity. *Annual Review of Anthropology*, 26, 291–312.

Ludtke, J., 1989. *Atlantic peeks—an ethnographic guide to the Portuguese-Speaking Islands*. Hanover: Christopher Publishing House.

Luz, S., 2007. The emergence of nationalist movements in Portuguese colonies in the 20th century: A case study of the Cape Verde islands and Guinea-Bissau. Dissertation, Kingston University.

Lynsap, M., 2000. Letter from Zambia. *The Nation*, February 14.

Maboreke, M., 1991. Women and law in post-independence Zimbabwe: Experiences and lessons. In Susan Bazilli, ed. *Putting women on the agenda*. Johannesburg: Ravan. 217 247.

MacDonald, T. & Gibson, J., 2007. *The global human right to health*. Abingdon, UK: Radcliffe.

Madison, D.S., 2005. *Critical ethnography: Method, ethics, and performance*. Thousand Oaks: Sage.

Maguire, P., 1987. *Doing participatory research: A feminist approach*. Amherst: Center for International Education.

Makwana, R., 2006. *Cancelling third world debt* [Online]. Available at: www.stwr.org/aid-debt-development [accessed October 2008].

Marques, M., Santos, R. & Araújo R., 2001. Ariadne's thread: Cape Verdean women in transnational webs. *Global Networks*, 1 (3), 283–307.

Marx, K. & Engels, F., 1998 [1845]. *The German ideology, including theses on Feuerbach*. Amherst, NY: Prometheus.

Maurice, F., 2001. Deconstructing gender: The case of Romania. In M. Hellinger & H. Bussmann, eds. *Gender across languages: The linguistic representation of women and men*. Amsterdam: John Benjamins. 229–254.

Máximo, S., 1998. *The batuque: A study of a female musical, literary performance from Cape Verde*. M.A. Thesis. University of London.

May, T., 2001. *Social research: Issues, methods and process*. London: Open University Press.

Maykovich, M.M., 1977. The difficulties of minority researchers in minority communities. *Journal of Social Issues*, 33, 108–109.

Mazur, J., 2000. Labor's new internationalism. *Foreign Affairs*, 79 (1), 79–93.

McGill, C., 2003. *Human traffic—sex, slaves and immigration*. London: Vision paperbacks.

McGowan, L., 2007. *Bailouts for bankers, burdens for women* [Online]. Available at: www.50years.org/factsheets [accessed May 10, 2007].

Mendoza, Z.S., 1998. Defining folklore: Mestizo and indigenous identities on the move. *Bull. Latin Am. Res*, 17 (2), 165–183.

Mendoza-Denton, N., 1999. Sociolinguistics and linguistic anthropology of US Latinos. *Annual Review of Anthropology*, 28, 375–395.

Merton, R., 1972. Insiders and outsiders: A chapter in the sociology of knowledge. *American Journal of Sociology*, 78, 9–48.

Merton, T., 1980. *Seeds of destruction*. New York: Farrar, Straus & Giroux.

Miles, M. & Huberman, A.M., 1994. *Qualitative data analysis: An expanded Sourcebook*, 2nd edition. Thousand Oaks: Sage.

Millennium Challenge Corporation, 2008. *Cape Verde compact progress: Highlights on the ground* [Online]. Available at: www.mcc.gov/countries [accessed on July 30, 2008].

Mohanty, C., 2003. *Feminism without borders: Decolonizing theory, practicing solidarity*. Durham: Duke University Press.

Moore, H L., 1988. *Feminism and anthropology*. Minneapolis: University of Minnesota.

Moran, E.,1982. The evolution of Cape Verde's agriculture. *African Economic History*, 11, 63–86.

Moser, C., 1993. *Gender planning and development theory: Practice and training*. London: Routledge.

Moser, C. & Moser A., 2005. Gender mainstreaming since Beijing. *Gender and development*, 13 (2), 11–23.

Mullin, M., 1994. *Africa in America*. Champaign: Illinois Press.

Munck, R., 2005. *Globalization and social exclusion*. Bloomfield: Kumerian.

Murray, D., 1998. Ebonics: A case study in language, power, and pedagogy. *TESOL Quarterly*, 32 (1), 144–46.

Naples, N., 1996. A feminist revisiting of the insider/outsider debate: The outsider phenomenon in rural Iowa. *Qualitative Sociology*, 19, 83–106.

Naples, N. & M. Desai, 2002. *Women's activism and globalization: Links in local struggles and transnational politics*. Routledge.

National Geographic, 2005. Washington, DC: National Geographic Society.

National Institute of Health, undated. Regulations and ethical guidelines [Online]. Available at: www.ohsr.od.nih.gov [accessed November 5, 2008].

Nationmaster, 2000. *Cape Verde* [Online]. Available at: www.nationmaster.com [accessed November 1, 2005].

New Economic Foundation (NEF), 2006. *Growth isn't working* [Online]. Available at: www.neweconomics.org [accessed November 23, 2008].

Nhlapo, T., 1991. Women's rights and the family in traditional and customary law. In Susan Bazilli, ed. *Putting women on the agenda.* Johannesburg: Raven. 111–123.

Novak, C., 1990. *Sharing the dance.* Madison: University of Wisconsin.

Ntshinga, T., 1996. Gender in Xhosa proverbs: Women's new expressions of emancipatory concerns. *Language Matters,* 27, 7–22.

Oakley, A. 1981. Interviewing women: A contradiction in terms. In H. Roberts, ed. *Doing feminist research.* London: Routledge. 30–61.

Obbo, C., 1980. *African women: Their struggle for economic independence.* London: Zed Books.

OSAGI, 2008. Gender mainstreaming [Online]. Available at: http://www.un.org/womenwatch/ [accessed November 10, 2008].

Palmberg, M., 2002. Expressing Cape Verde: *Morna, funana* and national identity. In M. Palmberg & A. Kierkegaard, eds. *Playing with identities in contemporary music in Africa.* Upsala: Nordiska Africainstitutet in cooperation with the Sibelius Museum/Department of Musicology, Abo Akademi University, Finland. 117–133.

Parker, I., 2005. *Qualitative psychology: Introducing radical research.* London: Routledge.

———, 2004. Discourse Analysis. In U. Flick, E. Kardoff, & I. Steinke, eds. *A companion to qualitative research.* London: Sage. 308–312.

Parks, J., 2006. *House passes Jubilee debt relief.* April 18 [Online]. Available at: www.blog.aflcio.org/2008/04/18 [accessed November 30, 2008].

Parpart, J.L., Connelly, M.P., & Barriteau, V.E., 2000. *Theoretical perspectives on gender and development.* Ottawa: International Development Research Center.

Payne, G., Williams, M., & Chamberlain, S., 2004. Methodological pluralism in British sociology. *Sociology,* 38 (1), 153–163.

PERCS, undated. *(Program for ethnographic research and community studies).* Elon University [Online]. Available at: http://org.elon.edu/percs [accessed September 4, 2008].

Petras, J. & Veltmeyer H., 2001. *Globalization unmasked: Imperialism in the 21st century.* Halifax: Fernwood.

Pike, K., 1971. Crucial questions in the development of tagmemics: The sixties and seventies. In Richard J. O'Brien, ed. *Report of the 22nd Annual Round Table Meeting of Linguistics and Language Studies,* 79–98. Monograph Series on Languages and Linguistics, 24. Washington, DC: Georgetown University Press.

Platt, J., 1981. On interviewing one's peers. *British Journal of Sociology,* 32, 75–91.

Potash, B., 1995. Women in the changing African family. In M.J. Hay & S. Stichter, eds. *African women South of the Sahara.* New York: Longman. 69–72.

Potts, L., 1990. *The world labor market: A history of migration.* London: Zed Books.

Powers, M. 1986. *Oglala women: Myth, ritual and reality.* Chicago: University of Chicago Press.

Prosser, J., 1992. Personal reflections on the use of photography in an ethnographic case study. *British Educational Research Journal*, 18 (4) 397–411.

Rabinow, P., 1977. *Reflections on fieldwork in Morocco.* Berkeley: University of California Press.

Rafter, N.H. & Stanko, E.A., 1984. Introduction. In N.H. Rafter & E.A. Stanko, eds. *Judge, lawyer, victim, thief: Women, gender roles and criminal justice.* Boston: Northeastern University. 1–26.

Rai, S., 2002. *Gender and the political economy of development from nationalism to globalization.* Cambridge: Polity Press.

Randall, B., 1991. *When is a pig a hog? A guide to confoundingly related English words.* New York: Galahad Books.

Randerson, J., 2006. World's richest 1% own 40% of all wealth, UN report discovers. *The Guardian*, December 6.

Razvani, S. & Miller, C., 1995. *From WID to GAD.* UN fourth conference on women, occasional paper #1. Geneva: Women's Research Institute.

Reinharz, S., 1992. *Feminist methods in social research.* New York: Oxford.

Reis, J.J., 2006. *Batuque*: African drumming and dance between repression and concession, Bahia, 1808–1855. In M. Gomez, ed. *Diasporic Africa.* New York: New York University Press. 45–63.

Reskin, B. & Padavic, I., 2001. Sex differences in moving up and taking charge. In L. Richardson, V. Taylor, & N. Whittier, eds. *Feminist frontiers V.* New York: McGraw-Hill. 253–262.

Robinson, C., 1981. Amilcar Cabral and the dialectic of Portuguese colonialism. *Radical America,* 15 (3), 39–57.

Scheuermaier, M., 2004. Economy. In M. Scheuermaier, ed. *Africa South of the Sahara.* London: Taylor & Francis. 189–191.

Schwartz, D., 1989. Visual ethnography: Using photography in qualitative research. *Qualitative Sociology*, 12 (2), 119–154.

Scott, K., 2008. *Change makers—Kum Kum Bhavnani* [Online]. Available at: www.rwmc.uoguelph.ca [accessed November 23, 2008].

Sen, G. & Grown, C., 1987. *Development, crises, and alternative visions.* New York: Monthly Review.

Shah, A., 2007. *PRSPs replace SAPs but still SAP the poor* [Online]. Available at: www.globalissues.org [accessed September 4, 2008].

———, 2001. Free trade and globalization [Online]. Available at: www.globalissues.org [accessed November 1, 2005].

Smith, I., March, C., & Mukhopadhyay, M., 1998. *A guide to gender-analysis frameworks (Oxfam skills and practice).* London: Oxfam.

Solomon, M., 1992. *We can even feel that we are poor, but we have a strong and rich spirit—learning from the lives and organization of the women of Tira Chapeau, Cape Verde.* Dissertation, University of Massachusetts.

Stacey, J., 1988. Can there be a feminist ethnography? *Women's Studies International Forum*, 11, 21–27.

Stanley, L. & Wise, S., 1983. *Breaking out: Feminist consciousness and feminist research.* London: Routledge.

Steady, F.C., 1981. The black woman cross-culturally: An overview. In F.C. Steady, ed. *The black woman cross-culturally*. Cambridge: Schenkman. 7–42.

Steinberg, L., Greenberger, E., Garduque, L., Ruggiero, M., and Vaux, A. , 1982. Effects of working on adolescent development. *Developmental Psychology*, 18, 385–395.

Stewart, E. & Bennett S., 1991. *American cultural patterns: A cross cultural perspective*. Yarmouth: Intercultural Press.

Stewart, F. & Wang, M., 2004. *Do PRSPs empower poor countries and disempower the World Bank, or is it the other way round?* QEH Working Paper Series—QEHWPS, Working Paper Number 108.

Strober, M., 1984. Toward a general theory of occupational segregation. In B. Reskin, ed. *Sex segregation in the workplace: Trends, explanations, remedies*. Washington, DC: National Academy.

Taylor, J. & Stewart, S., 1991. *Sexual and domestic violence: Help, recovery and action in Zimbabwe*. Harare: A.von Glenn & J. Taylor.

The Papiamentu Language, undated [Online]. Available at: www.umassd.edu/specialprograms/caboverde [accessed October 18, 2007].

Thomas, J., 1993. *Doing critical ethnography*. Thousand Oaks: Sage.

Thorne, B., Kramarae, C., & Henley, N., 1983. *Language, gender and society*. Rawley: Newbury House.

UNESCO, 2007. *Quiz* [Online]. Available at: www.unesco.org [accessed May 10, 2007].

———, 2004. *Slavery in Cape Verde*. [Online]. Available at: http://portal.unesco.org/ci/en/ev.php [accessed October 2008].

———, 2002. Universal Declaration on Cultural Diversity [Online] Available at: www.unesco.org [accessed November 23, 2008].

United Nations, 2008. *Human development report: 2007–2008* [Online]. Available at: http://hdrstats.undp.org [accessed July 10, 2008].

———, 2007. Convention on the Elimination of all Forms of Discrimination Against Women [Online]. Available at: http://www.un.org/womenwatch [accessed November 10, 2008].

———, 2006a. *Feminization of migration, remittances, migrants' rights, brain drain among issues, as population commission concludes debate* [Online]. Available at: www.un.org [accessed September 4, 2008].

———, 2006b. *Women's anti-discrimination committee chides delegation of Cape Verde over lateness in submitting first report* [Online]. Available at: www.un.org [accessed September 2, 2008].

United Nations Development Program, 1999. *Globalization with a human face: Human development report 1999*. New York: Oxford University Press.

United Nations International Children Fund (UNICEF), 1996. *The health of children less than five years of age in Cape Verde*. Ministry of Health, Cape Verde.

———, 1990. *The state of the world's children*. New York: Oxford University Press.

United Nations Population Fund, 2006. *New report calls on world leaders to protect human rights of female migrants* [Online]. Available at: www.unfpa.org [accessed September 4, 2008].

Urdang, S., 1989. *A revolution within a revolution: Women in Guinea-Bissau.* New York: Southern Africa Committee.

U.S. Department of State, 2008. Background note: Cape Verde [Online]. Available at: http://www.state.gov/ [accessed October 2008].

———, 2007. *Country reports on human rights practices* [Online] Available at: www.state.gov/ [accessed July 31, 2008].

U.S. State Department, 1989–1997. *Annual reports to the Committee on Foreign Affairs, U.S. House of Representatives and Committee on Foreign Relations, U.S. Senate, country reports on human rights, 1989–1997.* Washington, DC: U.S. State Department.

Van Dijk, T., Schiffrin, D., Tannen D., & Hamilton, H., 2003. *Handbook of discourse analysis.* London: Blackwell.

Wade, P., 2002. Music and the formation of the black identity in Columbia. *NACLA Report on the Americas*, 35 (6).

Wamboka, N.L., 2002. *Closest to the land, furthest from the cash: Poor women less likely to own land or have access to credit.* [Online]. Available at: www.africawoman.net [accessed November 25, 2008].

Werlin, L., 1995. *Jews in Cape Verde* [Online]. Available at: www.umassd.edu/specialprograms/caboverde [accessed November 1, 2005].

Westwood, S., 1984. *All day everyday.* London: Pluto.

Wilson, W., 1974. The new black sociology: Reflection on the insiders and outsiders controversy. In J.E. Blackwell & M. Janowitz, eds. *Black sociologists: Historical and contemporary perspectives.* Chicago: University of Chicago. 322–338.

Winford, D., undated. *Language in contact* [Online]. Available at: www.lsadc.org [accessed October 18, 2007].

Wodak, R. & Benke, G., 1997. Gender as a sociolinguistic variable. In F. Coulmas, ed. *The handbook of sociolinguistics.* Oxford: Blackwell. 127–150.

Wolfram, W., undated. *Sociolinguistics* [Online]. Available at: www.lsadc.org [accessed November 15, 2004].

World Bank, 2007. *Cape Verde: World Bank increases support for the growth and competitiveness project* [Online]. Available at: www.web.worldbank.org [accessed on July 30, 2008].

———, 2006. *Global economic prospects: Economic implications of remittances and Migration.* Washington, DC: World Bank.

World Bank Development Report, 2006. *Poverty in Cape Verde* [Online]. Available at: www.web.worldbank.org [accessed July 10, 2008].

———, 2004. *Making services work for poor people* [Online]. Available at: www.web.worldbank.org [accessed November 1, 2005].

Worth, S., 1980. Margaret Mead and the shift from "visual anthropology" to the "anthropology of visual communication." *Studies in Visual Communication,* 6 (1), 15–22.

Wright, C., Kumagai, F., & Bonney, N., 2001. Language and power in Japanese transplants in Scotland. *The Sociological Review,* 49 (2), 236–253.

Zavella, P., 1996. Feminist insider dilemmas: Constructing ethnic identity with Chicana informants. In D. Wolfe, ed. *Feminist dilemmas in fieldwork.* Boulder: Westview. 138–169.

———, 1989. The problematic relationship of feminist and Chicana studies. *Women's Studies,* 17, 25–36.

Index